Urban Re
Theory and

veying Series

Editi... ...ey

Emeritus Professor, Nottingham Polytechnic

Advanced Building Measurement, second edition, Ivor H. Seeley
Advanced Valuation, Diane Butler and David Richmond
An Introduction to Building Services Christopher A. Howard
Applied Valuation Diane Butler
Asset Valuation Michael Rayner
Building Economics, third edition Ivor H. Seeley
Building Maintenance, second edition Ivor H. Seeley
Building Quantities Explained, fourth edition Ivor H. Seeley
Building Surveys, Reports and Dilapidations Ivor H. Seeley
Building Technology, third edition Ivor H. Seeley
Civil Engineering Contract Administration and Control Ivor H. Seeley
Civil Engineering Quantities, fourth edition Ivor H. Seeley
Civil Engineering Specification, second edition Ivor H. Seeley
Computers and Quantity Surveyors A. J. Smith
Contract Planning and Contract Procedures B. Cooke
Contract Planning Case Studies B. Cooke
Environmental Science in Building, second edition R. McMullan
Housing Associations Helen Cope
Introduction to Valuation D. Richmond
Principles of Property Investment and Pricing W. D. Fraser
Quantity Surveying Practice Ivor H. Seeley
Structural Detailing P. Newton
Urban Land Economics and Public Policy, fourth edition P. N. Balchin, J. L. Kieve and G. H. Bull
Urban Renewal – Theory and Practice Chris Couch
1980 JCT Standard Form of Building Contract, second edition R. F. Fellows

Urban Renewal
Theory and Practice

Chris Couch

School of the Built Environment
Liverpool Polytechnic

First published 1990

Published by
MACMILLAN EDUCATION LTD
Houndmills, Basingstoke, Hampshire RG21 2XS
and London
Companies and representatives
throughout the world

Typeset by
Ponting–Green Publishing Services, London

British Library Cataloguing in Publication Data
Couch, Chris
Urban renewal : theory and practice. – (Building and surveying series).
1. Urban renewal
I. Title II. Series
711.4
ISBN 0–333–49644–2
ISBN 0–333–49645–0 Pbk

Contents

Preface

Urban renewal is of growing importance for two reasons. Firstly, as more of us live in towns and cities, urban areas become larger and older, so inevitably more and more renewal of the urban fabric has to take place. Thirty years ago it was possible to say that the vast majority of the buildings and infrastructure in British cities, outside city centres and inter-war clearance areas, had been subject to no renewal activity (replacement or refurbishment) whatsoever. Today most of the pre-1939 parts of all cities and much of the stock built in the 1950s and 1960s (especially 1960s local authority housing estates) have experienced some form of urban renewal: refurbishment, conversion or replacement. Secondly, there is growing concern about the constant expansion of towns and cities into their agricultural hinterlands, while large quantities of urban land and buildings are abandoned and left derelict. This lobby calls for resources to be directed towards the proper re-use or redevelopment of the existing urban fabric before further land is taken, usually irrevocably, into urban use.

In this field of activity, the built environment professions are increasingly taking on roles as redevelopers and 'implementation agents' in addition to their traditional advisory, design and plan-making functions, with a consequent blurring of the distinctiveness of each profession. Such developments have led to a need for all professionals, politicians and community groups involved in urban renewal to have a greater awareness of the economic and social context, as well as enhanced skills in urban design and the implementation of renewal schemes. To meet these needs we at Liverpool Polytechnic recently started a Masters degree programme in urban renewal. However, we soon discovered that while much excellent material had been written and published in the variety of fields that impinge upon urban renewal, there was a relative absence of books that sought to bring these strands together and provide an introduction to the subject.

Thus this book sets out to fill the gap and provide an introduction to the theory and practice of urban renewal. But this text is not entirely without opinions and conclusions, and a secondary aim is to put forward tentative views about some of the key issues currently facing policy makers.

Chris Couch

Acknowledgements

I have been grateful to many of my colleagues at Liverpool Polytechnic for their support and assistance over the months that it has taken to write this book. But I would like to acknowledge the specific contributions made by Reg Triplett, the Polytechnic's Architecture & Planning Librarian, for his help in finding source material; to my colleagues John Herson, Dennis Donnelly and Rob MacDonald for the ideas and inspiration that they provided and to Ralph Morton for his constructive comments on earlier drafts of this material. For many of the original notes from which the section on High Rise Housing was drawn I am indebted to Rakesh Lal. I would also like to put on record my debt to former colleagues and students, from whom I learned so much in the excellent but now defunct Department of Town and Country Planning at Liverpool Polytechnic.

Acknowledgement and thanks are also due to Professor Ivor Seeley, Editor of this series of books, for his guidance and helpful suggestions throughout the period of writing. Nevertheless the errors and omissions in this text are entirely my own responsibility. Lastly, and most importantly, special thanks must go to my wife Lynda and children: Daniel, Anna and Thomas, for putting up with my long evenings spent at the word-processor in the back room.

1 Introduction

Urban areas are never static, they are constantly changing: either expanding, contracting or undergoing internal restructuring in response to economic and social pressures. In the developed economies of Western Europe the pattern of urbanisation and the structure of towns and cities has been laid down over many centuries. Population growth no longer exerts the pressures it did during the Industrial Revolution or following the great shifts of peoples that occurred after the two world wars. In this context much of the change that affects urban areas results from the gradual spatial and sectoral adjustments to economic activity and the movement of population between or within existing urban areas.

These spatial and sectoral changes in demand for land and buildings lead to the intensification of use in some areas, a reduction of density in others, in some cases to refurbishment and perhaps a change in the use of a building, in another case to demolition and reconstruction, and in a few cases to the abandonment of buildings, vacancy and dereliction. Furthermore, there will be public utilities, transportation infrastructure and social facilities to be provided, adapted, expanded, contracted or replaced in response to these changing demands. Inasmuch as these changes affect the physical structure and fabric of urban areas it is regarded here as a process which we shall call 'urban renewal'. In other words, for our purposes, urban renewal is seen as the physical change, or change in the use or intensity of use of land and buildings, that is the inevitable outcome of the action of economic and social forces upon urban areas.

Through the historical development of urban areas since the Industrial Revolution it is possible to detect a number of types or categories of urban renewal. The Industrial Revolution itself led to a reorganisation and expansion of many urban areas. As towns and cities expanded,so competition for the best locations increased; land and buildings changed use and densities of development and occupancy increased. A comparison of land use and density in any major British city between say, the end of the eighteenth century and the middle of the nineteenth century, will show considerable evidence of this restucturing or urban renewal. One of the features of this kind of renewal which, of course, continues right up to the present day, is that it is a very powerful force for urban change and predominantly a market-led process.

Since the middle of the nineteenth century the state has increasingly made efforts to intervene in the process of urban change and renewal in order to achieve the social objectives that are not met by the workings of the land and property markets. From that time the state began to control the design of streets and utility networks and to set minimum standards for housing; to take powers to require the removal of slum property and even (in a very limited way at first) to build replacement housing itself. Gradually, the state established a role for itself in defraying some of the social costs associated with urban development and renewal.

The state has also become involved in urban renewal in another way. This time not in the field of social expenses but in overcoming market imperfections and undertaking social physical capital investment. Over the years it has become apparent that some aspects of urban renewal are not easily handled by the market. There are sometimes major infrastructure investments required to unlock potential sources of profitable property development; but the supply of physical urban infrastructure is seldom profitable and so its provision has become one of the tasks of the state. Urban highways, car parking facilities and public transportation systems are examples of the unprofitable but essential urban framework that is required before profitable property development can take place.

Sometimes there are new or uncertain markets where state action is required before investor confidence can be established, such as in overcoming the 'prisoners' dilemma' in housing improvement, or in reducing the risk associated with private inner city speculative housing investment. Sometimes there are economic inefficiencies, such as the existence of excessive vacancy in the urban land market, which require state expenditures to reduce redevelopment costs to a level that will induce the private market to function.

Thus the processes that have led to the renewal of urban areas in Britain can be subdivided into three types of activity:

(i) market-led renewal and restructuring without state intervention;
(ii) urban renewal which takes the form of social expenses regarded by the state as necessary for social harmony and well-being (mainly in the fields of housing, public health and environmental policies);
(iii) social physical capital investment and state regulation to facilitate profitable private sector property development or redevelopment (such as the subsidies provided for derelict land reclamation, infrastructure and urban public transport provision).

A distinction should be made between this process of essentially physical change, which is referred to here as urban renewal, and the wider process of 'urban regeneration', in which the state or local community is seeking to bring back investment, employment and consumption and enhance the quality of life within an urban area. This book is not so wide in its scope or ambitions: it does not present a comprehensive analysis of the inner city problems or the urban regeneration process, although it does deal with a number of aspects of regeneration where they most obviously have physical dimensions or implications.

The purpose of this book is to provide an introduction to relevant theory and practical experience of urban renewal for students, and for people and organisations, whether from the state or local communities who wish to intervene in the 'natural' market-led process of physical urban renewal in order to modify the outcomes of the process in some way. This modification may be to improve directly the social well-being or quality of life of a local community; it may be to assist local economic development and job creation; or it may be to enhance the aesthetic value or reduce the environmental costs of change. Whatever the reason for intervention, there are certain prerequisites for action: it is important to know the historical context of change; the economic possibilities and limitations; the social implications of action; the appropriate organisational structure and managerial approach to adopt; and the physical opportunities and constraints presented by the circumstances. In this book each of these requirements is introduced and discussed in turn.

The structure of the book is quite straightforward. Chapter 2 offers an overview of the historical evolution of urban renewal; Chapters 3 to 6 consider aspects of relevant urban economic, social, managerial and design theories. In Chapter 7 this knowledge is applied to a discussion of current urban renewal practice. From this, conclusions are drawn in Chapter 8 on the nature of the lessons to be learned, and the importance of certain considerations, and some 'rules of thumb'are suggested for good practice.

The historical review in Chapter 2 is not intended to be comprehensive but to draw out a number of key events and important lessons for modern urban renewal practice. This review begins with the Industrial Revolution. Naturally, there were many years of urban history and substantial achievements in urban planning and design before this period, but the amount of urban renewal, as distinct from urban expansion, that occurred before the Industrial Revolution was limited and the period 1750–1800 represents some kind of break-point after which there was a significant change in the amount and nature of urbanisation and urban restructuring in Britain.

This historical discussion examines the nature of the market pressures for urban renewal in the nineteenth and twentieth centuries and the urban economic, social and physical consequences. Consideration is given to the beginnings of and reasons for state intervention in the renewal process and to key developments in policy, such as the beginnings of slum clearance, the introduction of land use planning, the moves towards and then away from high-rise housing, the shift towards area improvement policy and the development of state intervention in wider questions of urban regeneration and inner city policy. Three key influences on the changing economic context of urban renewal are identified in the chapter: the crucial role played by developments in transportation systems; the effects of manufacturing decline and service sector growth in changing the nature and location of demands for urban space; and thirdly, consequent upon the first two phenomena, the increasing divergence in the economic conditions under which urban renewal occurs between central and peripheral regions.

The consideration of economic theory in Chapter 3 begins with an analysis of the demand for urban construction and urban renewal before focusing on the particular characteristics of housing demand. This is followed by a complementary review of the supply side: the building and development sectors. Attention then turns to questions of the economic life of buildings, the timing of redevelopment and refurbishment decisions and the choices available to building owners, including the possibility of vacancy and dereliction. Some views about the dynamics of urban renewal markets having been established, the issues of justification and methods of state intervention are raised and debated. The chapter concludes with a discussion about the nature of urban industrial change and the economics of urban regeneration.

Chapter 4, on the social aspects of urban renewal, considers the predominant trends in urban population and household structure and movement; the nature and importance of communities and neighbourhoods, and the interaction between urban renewal processes and vulnerable groups: women, racial minorities and the elderly.

Chapter 5, on management and organisational theory, deals with planning and decision-making processes and styles and examines the scope for transferring theories developed for the better management of private businesses to the process of public intervention in urban renewal. Particular attention is paid to the role of community participation in renewal decisions.

The development of urban design theory in Chapter 6 is traced from the early work of Kevin Lynch and British writers such as Sharp and Cullen through to current ideas about conservation and responsive environments. Connections are made between physical urban forms and urban economic forces. The chapter ends with consideration of some of the emerging concerns with longer term ecological and environmental objectives in urban design.

Chapter 7 is the largest in the book and deals with current practice in urban renewal. It is structured around the three possible types of state intervention: regulation; spending and taxation; and organisational changes including the establishment of new agencies. In each section policy instruments are described and critically discussed. Where appropriate case studies and examples are presented so as to illustrate and ciarify the issues being raised. Land use planning, conservation and preservation policies and land registers are considered under the heading of 'regulation'. The majority of current government urban renewal policies fall under the heading 'spending and taxation': housing renewal subsidies, derelict land grants, the urban programme, enterprise zones (although these include some regulation), and industrial and commercial improvement areas. 'New agencies and organisational changes' includes urban development corporations, other public agencies, private and charitable agencies and the cooperative movement.

In the concluding Chapter 8 attention is drawn to the close interaction between urban social, economic and political processes, urban design and the management processes in policy implementation and the importance of developing understanding and linkages between these elements in urban renewal. The tension between short run market economic pressures and longer run social needs is considered and hope is seen in the way that discussion of social costs and benefits is now broadening to include greater acceptance of health, energy and ecological objectives. It is also apparent from the history of urban renewal, from management theory and from experiences in current practice that local community participation is another essential component of good practice in urban renewal.

For the most part the book is factual or interpretive of earlier work but on occasions does include the author's own views or conclusions. The places where this occurs are prefectly clear in the text and no apology is made for their inclusion. At the very least they will provide readers with a starting point for debate and discussion.

2 Aspects of the Historical Development of Urban Renewal

2.1 The Industrial Revolution, Housing and the Construction Industry

According to Hobsbawn the Industrial Revolution in Britain was:

> not merely an acceleration of economic growth, but an acceleration of growth because of, and through, economic and social transformation,
> (Hobsbawn, 1968, p. 34)

and this was as true for the scale and nature of city development as it was for any other aspect of British life. During the second half of the eighteenth century and the first forty years of the nineteenth century the British economy experienced:

(i) a rapid acceleration in investment leading to cumulative growth (i.e. it was essentially a 'capital building' period which included not only investment in plant and machinery but also a massive concomitant expenditure on urban infrastructure: factories, housing and transportation and utility networks);
(ii) rapid innovation in technical and social relations including not only the new relationships between employers and employees required by the factory system but also a social restructuring and a new social order within cities including ultimately a much expanded 'middle' class that proved highly influential in shaping future processes of urban development.

The construction and development industries made major contributions to this process of 'industrial revolution' as:

(i) Plant and machinery had generally to be accommodated within buildings and workers had to be housed near factories. Urban development was thus a key process in facilitating capital accumulation.
(ii) The construction process employed about 17,000 workers by 1800, or 3.5 per cent of the workforce, by 1841 the industry employed around 4.7 per cent of the workforce and housing output alone had risen to 8.2 per cent of GNP (all figures calculated from data in Deane & Cole (1962)). Construction was therefore increasingly important to the national economy, and housebuilding in particular

achieved greater productivity (in terms of GNP added per worker) than the industrial average.

(iii) The construction industry was a major consumer of the output of growing, mechanising industries such as ironwork, glass, brick and tile. These materials often had to be transported considerable distances, so making the urban development process one of the most important users of the emerging transportation system. (Powell, 1980, p.27)

Between 1760 and 1860 the country's capital stock increased dramatically with the per capita value of industrial and commercial capital (including buildings) rising from £7.40 to £50.90 (+588 per cent), while the per capita value of residential and social capital (mainly housing) rose by only 141 per cent from £16.20 to £39.10 (Feinstein 1981, p136). In order to achieve this increased rate of investment it was necessary to increase the savings ratio. This was achieved both through ploughing an increased proportion of profits back into capital investment, and holding wages and social expenditure down so as to create 'enforced savings' that could be transferred into industrial investment. The general surplus of labour and the relative power of employers during this period contributed to the success of this process and were important reasons why the quantity and quality of industrial and commercial building and transportation infrastructure accelerated so far ahead of the development of urban housing and social infrastructure and why they led to the apparent deterioration in urban housing conditions during the early nineteenth century.

The growth in industrial and commercial activity led to increased urbanisation. This in turn led to scale economies within enterprises and agglomeration economies within cities. Both factors led to further urban expansion which in turn increased the demand for construction activity. Much of the expansion of these urban areas resulted from in-migration rather than natural change as increasing rural poverty encouraged migration towards the cities in search of work. Frequently these newcomers helped to construct the very cities to which they had migrated:

> the field (casual unskilled building labouring) was a favourite entry point into employment for migrants.
>
> (Powell, 1980, p.32)

Prior to 1750 population growth and movement were slight. Housebuilding activity was at a very low level and mainly confined to replacement of the existing stock, probably around 1 per cent or 10,000 dwellings a year nationwide. Then Ashton suggests that:

> the pace increased during the following three decades, and that it was rapid in the last twenty years of the century.
>
> (Ashton, 1972, p.97)

It seems likely that housebuilding increased to around 24,000 dwellings a year by the turn of the century and 60–70,000 dwellings a year by the 1830s. In each period perhaps half of this construction would have been for replacement housing (housing renewal). The output of industrial and commercial building and civil engineering works would have been accelerating at a much faster rate over this period.

While most non-housing construction would have taken place within the market, prior to 1750 a fairly high proportion of housing output would have been built by workers for their own consumption or by employers (especially in rural areas). In neither case would housing have been provided as a commercial operation. Only a small proportion of dwellings, possibly less than 5,000 a year mainly in London, would have been supplied under market conditions. However, most of the subsequent increase in output was spatially concentrated in urban housing markets, built with wage labour and purchased materials on land acquired for the purpose and then sold or rented for profit. Thus housebuilding came almost entirely within the capitalist market economy and was increasingly provided by specialist building firms.

2.2 Market Forces and Urban Renewal in the Nineteenth Century

The Industrial Revolution and the imperatives of capitalism had profound effects on the development of urban areas and upon urban renewal. Five distinct types of urbanisation can be identified:

(1) Primary urbanisation associated with the creation of new workplaces providing space for basic economic activity (i.e. industries that export goods to other areas) and housing for the workers and their families.
(ii) Secondary urbanisation associated with the development of non-basic economic activity (such as service employment) and housing for these workers and their families.
(iii) An upward spiral of tertiary urbanisation based upon the larger markets, agglomeration economies and comparative advantages that emerged with the growth of these industrial and commercial centres.
(iv) The growth of suburbs, spas and seaside resorts based upon the growing wealth and surplus incomes of the expanding middle class.
(v) The growth of London which, because it is so different in scale and historical experience from other cities, has to be identified as a separate type. London expanded rapidly during the nineteenth century as the expanding national economy required ever greater and more sophisticated financial services, which, for historical and practical reasons were heavily concentrated in the City of London. This growing and more complex economy required more and better

> central government; London also had its own industrial base and agglomeration economies far surpassing those of any other city, and a phenomenal concentration of middle class and aristocratic wealth.

Within these urbanisation processes, market forces were beginning to lead to significant internal urban restructuring and the renewal of large areas. As the size of urban areas grew so did competition for the best-located sites, usually the most central sites. The effect of this competition was to bid up the price of land. As land prices rose so it was necessary to increase the intensity of land use, whether for production or housing purposes, so as to maximise profits. Sites that were being used at sub-optimal intensity (i.e. low building densities) would tend to be cleared and rebuilt at higher densities. This applied in industrial and commercial areas where small scale workshops might be replaced by larger multi-storey premises; and in housing areas where early industrial cottages might be replaced by more profitable terraced and courtyard developments; and also between housing and other uses where the more profitable industrial or commercial uses might replace housing on better located more central sites. Figures 2.1 and 2.2 show the increases in urban densities and changes of use occurring in Liverpool in the nineteenth century.

A second major force for urban renewal was the expansion of the service sector. As trade and industry developed, an ever increasing range and sophistication of economic infrastructure was required: trade required banking facilities, insurance, transportation agents, legal services and a host of other ancillary activities. It also required more and better central and local government regulation and support. These changes led to a growth in office employment that began in earnest in the eighteenth century mercantile period, accelerated during the nineteenth century and has continued ever since. Many offices, because of their close economic linkages with one another and the need for personal contacts, desired the most accessible central area locations. Because of their very high levels of profit per unit of area compared with other land using activities they tended to be able to outbid and replace other uses in such locations.

Another and not unrelated trend, during the nineteenth century, was the growth of the middle class and the rise in their aggregate spending power. Much of this rising income was spent on housing, for as Burnett has said:

> This was the most family-conscious and home-centred generation to have emerged in English history,
>
> (Burnett, 1978, p.95)

so leading to an acceleration in suburbanisation. However, a proportion of this income was spent in other ways, notably: (i) in taxation and local rates, some of which was used for civic improvements, and (ii) in spending on luxury goods, clothing and other household items. The increasing

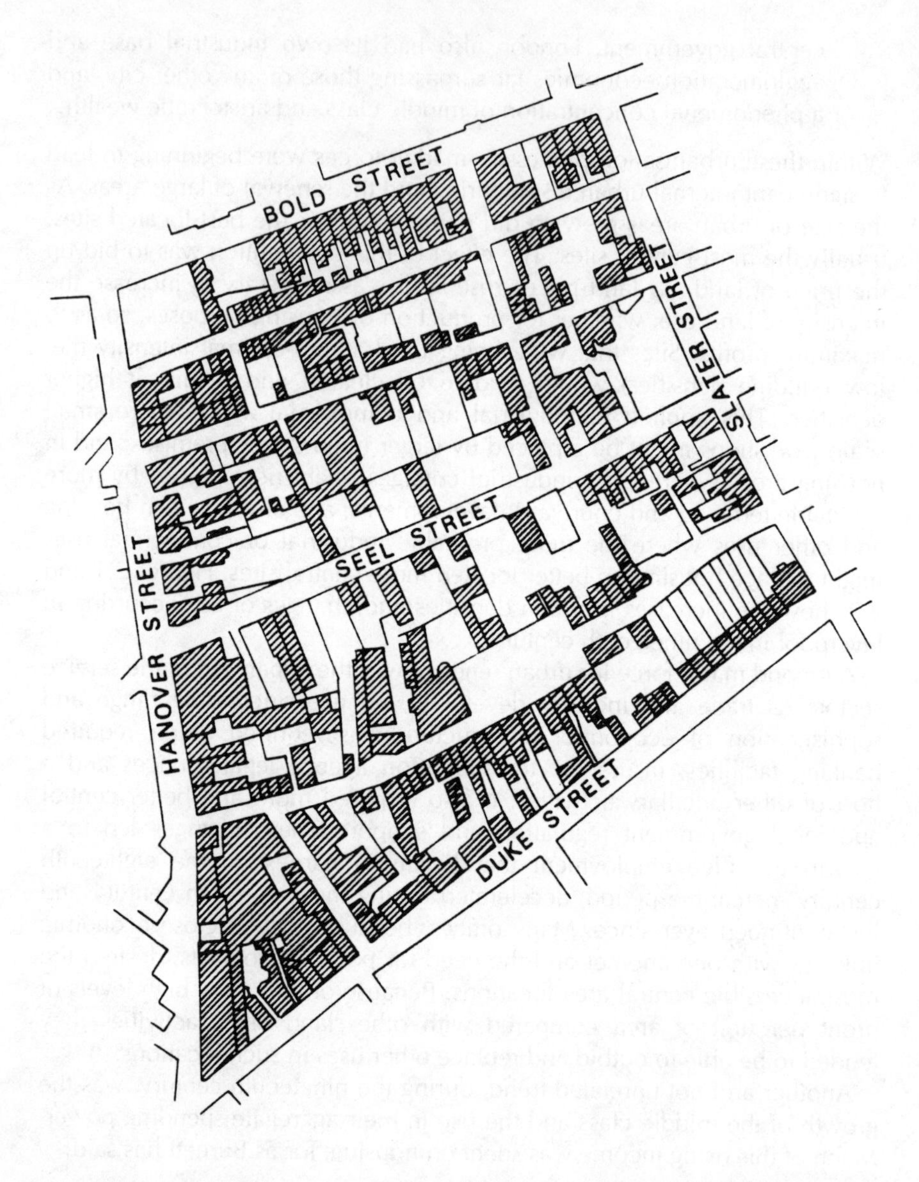

Figure 2.1 The pattern of urban development, central Liverpool 1803

volume of sales gradually led to the establishment of shops in recognisable 'shopping centres' within the Victorian city. The profit obtainable per unit of area of space from retailing was greater than almost any other land use except offices and, because of the need for accessibility (especially for comparison shopping) tended to gravitate towards the city centres.

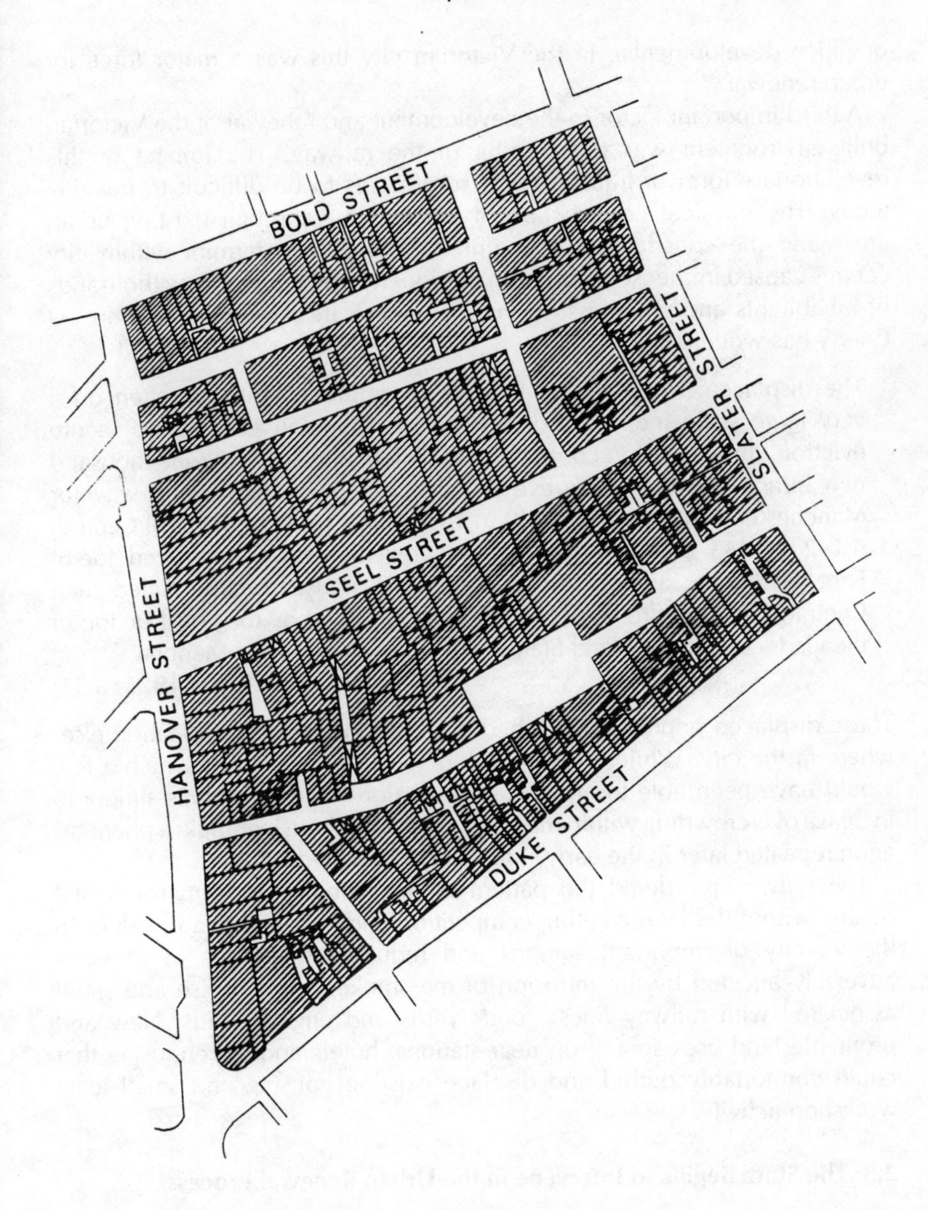

Figure 2.2 The pattern of urban development, central Liverpool 1895

So there were two land uses, offices and shops, which were frequently complementary users of the same parcel of land, developing in central areas and displacing the previously established but less profitable land uses (which might have been other activities or simply less intensive shopping or office developments). In the Victorian city this was a major force for urban renewal.

A third important factor in the development and renewal of the Victorian built environment was the coming of the railway. The impact of this revolutionary form of transport was so great as to be difficult to imagine today. The physical construction of the railways through existing urban areas and the construction of mainline stations and termini within city centres caused immense destruction of property and forced many thousands of inhabitants and countless businesses to seek fresh accommodation. As Cherry has written:

> The displacement of people for railway improvement was often on a very large scale indeed, and, because it was relatively easy to secure eviction, this expulsion could be both violent and sudden. One thousand two hundred and seventy-five people in 255 cottages made way for Manchester Central Station; 540 in 135 houses for the Liverpool Central; 6,142 in 443 of the larger Glasgow tenements were required for St Enoch's Station; and 2,178 people in 141 houses for the Caledonian Central. Eight hundred families were dispossessed by the construction of the viaduct linking Central Station, Newcastle, to nearby Manors.
>
> (Cherry, 1972, p.41)

These displaced populations pushed up demand for accommodation elsewhere in the city. While this resulted in some new building, which a few would have been able to afford, for the majority the effect was simply to increase overcrowding within the remainder of the housing stock: a phenomenon repeated later in the early slum clearance schemes.

The railway reordered the pattern of accessibility and environmental quality within the city, creating competition and enhancing land values in the vicinity of important stations and reducing the value of property adversely affected by the intrusion of the smoke, grime, noise and smell associated with railway lines, goods yards and engine sheds. New and profitable land uses sprang up near stations: hotels and warehousing that could comfortably outbid and displace existing housing and small-scale workshop activity.

2.3 The State Begins to Intervene in the Urban Renewal Process.

Apart from indirect involvement, for example in the railway building process, there had been little significant state concern with urban renewal until the 1860s. While there was perhaps little perceived reason to involve the machinery of the state in the urbanisation process or the growth of city centres, the failure of the state to react to the appalling, and in many cases deteriorating, condition of the housing stock and working class living conditions, needs some discussion.

There were a number of reasons for the slow rate of housing renewal and improvement during the first half of the nineteenth century. Undoubtedly the most important reason was the low level of real working class incomes caused by their lack of bargaining power and their consequent inability to purchase any better housing conditions in the market-dominated economy of the time. Other reasons all concern inadequacies within the structure of society or the state. Gaudie (1974) puts forward a number of propositions and these are summarised here.

(i) There was a low level of knowledge about the nature of the housing problem and about the appropriateness of solutions. Few middle class people would have had reason to enter working class housing areas and there was hardly any 'media coverage of the issue' as we would know it today. Knowledge was based upon rumour and supposition. Official statistics on housing conditions were beginning to appear during the century: first with the publication of decennial censuses after 1801; with the registration of births and deaths, mortality statistics and the reports of the Poor Law Commissioners from the 1830s; and with the growth of unoffical statistics collected by insurance companies and scientific societies. It was not until the late 1840s that the causes of cholera, typhoid and related diseases began to be understood and the importance of clean water supply, proper drainage and building ventilation appreciated.

(ii) The dominant economic philosophy of minimum state intervention in the running of the market, at least as far as the internal economy of the country was concerned, made the government reluctant to impose regulations on the workings of any market sector and housing was seen to be little different from any other commodity in this respect. There had been a long British tradition of the sanctity of private real property (i.e. landed property) so that what any landowner or landlord chose to do with his property tended to be regarded as a private matter subject only to the normal common law of remedies of contract and tort.

(iii) The power and leverage of the working class were limited. In the early part of the century the working class had little political power: it had no vote and had very little industrial muscle due to the absence of many large scale concentrations of employment and repressive state attitudes to the development of trades unions. Common law remedies against landlords for breach of contract or the torts of negligence or nuisance were of little value in slum districts where few could read, write or afford the costs of legal action and where many could easily be intimidated by a landlord or landowner forced to defend such an action.

(iv) The state system at local level was rudimentary and quite inadequate for the task of regulating and intervening in urban development. While borough councils existed in many of the older towns, much of the new industrial and residential development of the early ninteenth century was taking place outside these areas under the most elementary local government systems. The powers and duties of authorities were limited, the numbers and quality of staff often poor and corruption all too commonplace. A start was made on the reform and development of local government under the 1835 Municipal Corporations Act but outside London it was as late as the 1870s before a modern and effective system of local administration was installed.

It was left to the goodwill of industrialists such as Sir Titus Salt in Bradford (founder of Saltaire) and the Wilson brothers of Price's Candle Company (founders of Bromborough Pool Village) to initiate and subsidise good quality housing for their workers. But employer owned housing has never been a major feature of the British housing system and the quantitative contribution made by such housing was small. More significant in terms of the later development of housing policy was the establishment of the early philanthropic housing societies and trusts, such as the Peabody Trust (1862) and the Improved Industrial Dwellings Company (1863). While the total contribution of these societies was minimal and their failings were considerable they were important for two reasons: they were instrumental in bringing the slum housing problem to public attention and they formed the basis for the development of the modern housing association movement.

It was only as these impediments were gradually overcome during the century that the state began to develop and refine its role in urban development and renewal processes.

There had been major legislation under the 1848 Public Health Act but that was predominantly concerned with new building. However, by the 1860s there did seem to be real civic concern with the condition of existing buildings and particularly with the condition of housing within working class areas. In Liverpool, for example, the city council, having earlier banned the construction of cellar dwellings and courtyard housing began to build new tenement blocks on cleared sites as models of good and economical practice for the private sector to follow. Their first venture was at St Martin's Cottages in the Everton district. The idea was to prepare a model scheme to show what could be done and then to auction off the land. However,

> considerable apathy was shown amongst potential buyers and in October 1866 the Council decided on the recommendation of the Health Committee, to build the dwellings itself.
>
> (Sutcliffe, 1974, p.167)

Thus, more or less by accident, was born the first local authority replacement housing to be built on a site cleared of former slum property. The first national legislation concerned with the removal of slum housing came with the Torrens (1868) and Cross (1875) Acts.

When William Torrens introduced his Bill in 1868 it was designed to permit local authorities to require the clearance of streets of slum property and to build replacement housing themselves. Such proposals were controversial at the time and during its passage through Parliament the clause providing for replacement building was dropped and the word premises was substituted for the word street so that the Artisans and Labourers Dwellings Act, as it was officially known, applied to individual dwellings rather than whole insanitary areas. Local authorities were given permission to levy a twopenny rate for the purposes of the Act. However, what had started as a radical idea had been effectively reduced by the forces of conservatism to an ineffective policy and an administrative nightmare. Some authorities did make use of powers available under the Act to force through clearance programmes but the effect was often the reverse of what had been intended:

> Demolitions in the area of the Mint (in the City of London) under the Torrens Act caused local rents to double...
>
> (Burnett, 1978, p.149)

That clearance alone, without subsidised rebuilding, would lead to a worsening of working class living conditions was well appreciated by many authorities. Thus, taking the view that an unfit dwelling was better than no dwelling at all, many Medical Officers of Health were reluctant to declare property to be insanitary. Further, where dwellings had been declared insanitary, owners had the possibility of carrying out repairs and improvements instead of facing demolition without compensation: many chose to make the minimum necessary changes in order to continue earning rents.

The problem of dealing with slums on a dwelling by dwelling basis was resolved under the Artisans and Labourers Dwellings Improvement Act 1875 (The Cross Act) when local authorities were given powers to deal with insanitary areas. The Act also went some way towards recognising that rehousing was a public responsibility by giving local authorities power

> to draw up schemes for the improvement of slum areas, carrying out the street planning, paving and sewering of the land itself and granting or leasing the land to persons who would build upon it under conditions imposed by the authority.
>
> (Gaudie, 1974, p.276)

But these 'rebuilders' still received no housing subsidy, other than that coming from local rates, so rents, even those charged by the philanthropic societies, were well beyond the means of the majority of the dispossessed tenants of these areas. Also, of course, there was an inevitable time-lag before rebuilding on any cleared site so the immediate effect was again to reduce the total amount of cheap working class housing available. The obvious ineffectiveness of the legislation, together with the absence of subsidy, the apathy of local authorities and the power of vested interests meant that by 1884 only nine towns in the country had begun improvement schemes under the Act. As Gaudie has observed:

> The Improvement Acts of Torrens and Cross, although they were pioneering attempts at lifting the condition of the people, in fact increased the misery of cities. By allowing and encouraging the demolition of homes without a rehousing policy they inevitably increased overcrowding and the ills attendant upon it.....They did make the policing of the cities easier because anti-social elements were dispersed and took time to form again. They did not make homes of the people more comfortable or more abundantly available....Their real intention was to make cities pleasanter in appearance by removing the worst eye-sores among the slums and safer for the middle classes to walk in.
>
> (Gaudie, 1974, p.267)

In spite of the importance of these Acts as legislative landmarks, for they represented major intrusions into the former rights of property owners, the scale of their impact on housing renewal and urban change was limited. Indeed, urban renewal activity up to World War 1 was almost wholly the outcome of the interplay of market forces and was only marginally influenced by these state interventions. Figure 2.3 gives a stylised view of the processes of urban renewal in the nineteenth century.

2.4 The Inter-war Years

Slum clearance and housing improvement ceased with the outbreak of the World War 1. Four years of virtual inactivity in housebuilding left the country in 1918 with an acute housing shortage and a substantial backlog of repairs. The political necessity of the time was to deal with the housing shortage and a series of Housing Acts (Addison Act 1919, Chamberlain Act 1923 and Wheatley Act 1924) were all in their different ways intended to stimulate new housing supply. During the 1920s the condition of the existing dwelling stock was a relatively neglected issue with only a few thousand of the worst slums being cleared during the decade. Furthermore the effect of the rent controls imposed by the 1915 Rent and Mortgage Restriction Act and subsequent legislation was to deter landlords from carrying out increasingly unprofitable housing maintenance and repairs.

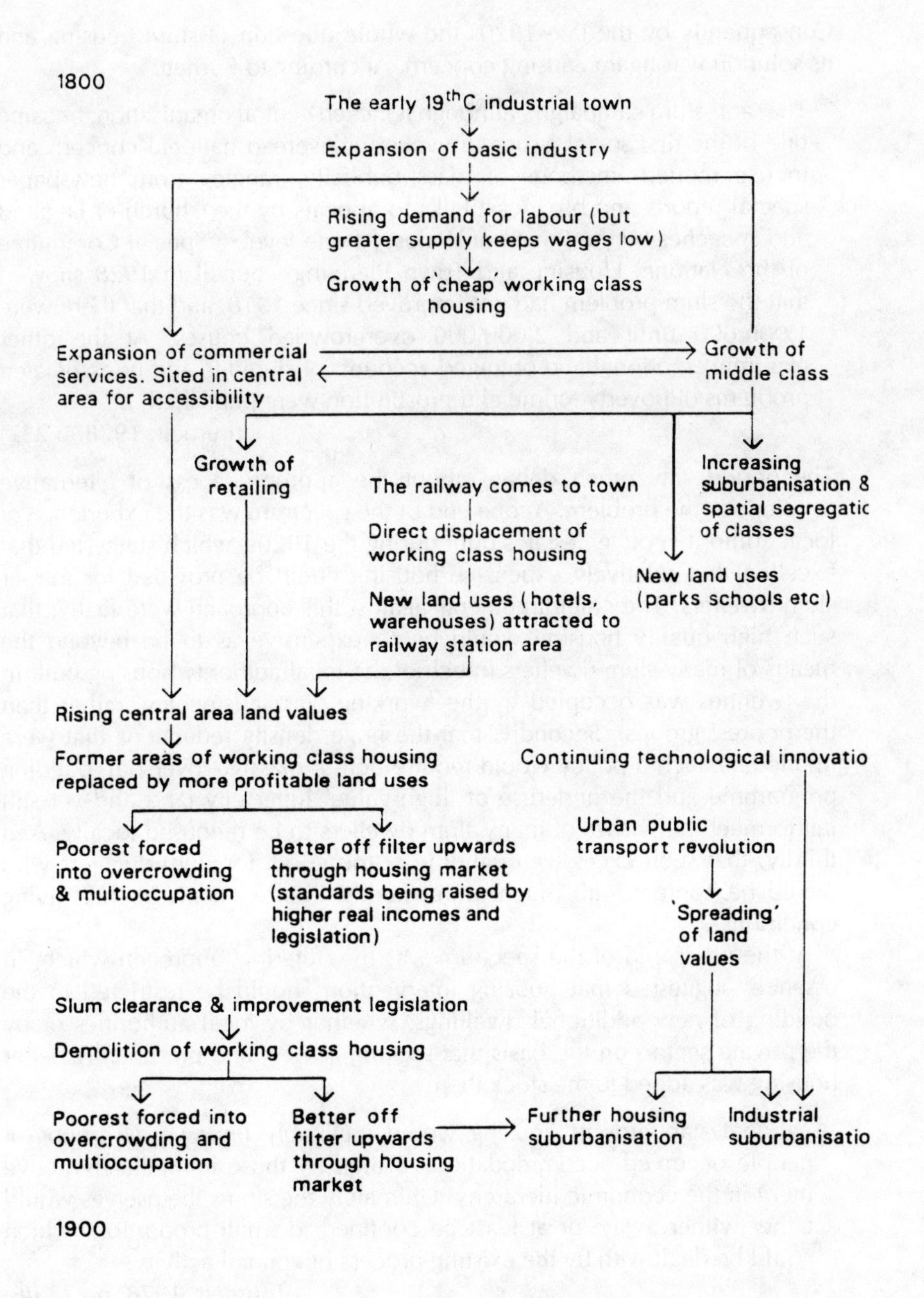

Figure 2.3 Urban renewal processes in the nineteenth century – a stylised view

Consequently by the late 1920s the whole question of slum housing and its solution was again causing concern. According to Burnett:

> The 'anti-slum campaign', although it lacked central organisation, became one of the first social issues to evoke widespread national concern and to use modern methods of mass publicity, ranging from newspaper special reports and broadcast talks to appeals by the Church of England and speeches by the Prince of Wales. At one level, a Special Committee of the National Housing and Town Planning Council in 1928 showed that the slum problem had not improved since 1918, and that there were 1,000,000 unfit and 2,000,000 overcrowded houses. At the other extreme, emotional and outraged accounts of slum life and its associated problems of poverty, crime and prostitution were published....
>
> (Burnett, 1978, p.237)

The period saw much debate about the appropriateness of alternative solutions to the problem. At one end of the spectrum was the experience of local authority cottage estates built during the 1920s, which suggested that excellent but relatively expensive housing could be provided for former slum dwellers. The chief arguments against this approach were firstly, that such high quality housing would be so expensive as to be beyond the means of many slum dwellers (much of the local authority housing built in the twenties was occupied by the 'working class aristocracy' rather than the poorest groups). Secondly, that the huge density reductions that were implied by such a policy would require both a massive 'overspill' building programme and the underuse of 'high-value' inner city land and yet still fail to meet the wishes of many slum dwellers to be rehoused locally. And thirdly, that such excessive quality was 'too good' for slum dwellers who would be content with much more modest improvements in their living conditions.

At the other end of the spectrum was the 'filtering' approach which, in essence, suggested that housing intervention should be restricted to the building of new additional dwellings, whether by local authorities or by the private sector, on the basis that so long as new and presumably better housing was added to the stock then:

> national standards of housing would inevitably improve as groups of people occupied accommodation vacated by those immediately above them in the economic hierarchy. Ultimately the slums themselves would either wither away, or at least be confined to small proportions which could be dealt with by the existing process of council activity.
>
> (Burnett, 1978, pp.235/6)

In the circumstances of the inter-war years, with household formation rates exceeding housing stock increases and with real barriers to working class mobility, filtering alone would not have been an effective mechanism for mass housing improvement. Between these two views was a third idea: the concept of the 'minimum standard' dwelling. It was argued that by exploiting economies of large scale production; reducing the size and specification of housing and increasing densities, more and cheaper accommodation could be built, thus better meeting working class needs both in terms of quantity and price.

This was a powerful argument and in this period, and on more than one subsequent occasion, could be said to have won the day. For the solution to the slum problem that was finally implemented in the 1930s was not to replace slums with cottage estates nor to rely solely upon filtering but to replace slums with a combination of walk-up tenement flats on the cleared sites and some overspill cottage estates. The tenements were to offer less space and higher densities than previous inter-war council housing schemes and the supposed, though largely erroneous, cost savings of building flats rather than houses. Even the overspill cottage estates of the 1930s were built to lower standards than previous local authority housing. The legislation that enabled implementation of this solution was the 1930 (Greenwood) Housing Act. Local authorities were required to submit five year plans for slum clearance and for the first time the Act:

> introduced an Exchequer subsidy specifically for slum clearance and, importantly, related the subsidy to the numbers of people displaced and rehoused: the intention was to prevent the pre-war practice of demolition by local authorities without replacement. Also by basing the subsidy on people rather than houses, it would make it easier for councils to deal with the problem of large, poor families, since the subsidy would increase with the size of the family rehoused.
>
> (Burnett, 1978, p.237)

Local authorities were free to adopt any pricing policy they chose so long as rents accorded with what tenants could reasonably be expected to pay. By 1933 when the scheme finally came into force, government efforts in the housing field had been reduced to an almost exclusive concentration on tackling the slum problem, and local authority building for general needs was virtually abandoned. By 1939 it was estimated that almost 500,000 slum houses still required demolition (almost certainly an underestimate) but that 245,000 had been cleared with 255,000 dwellings built in replacement: compared with recent performance in the late eighties this was quite a reasonable performance!

It was the amount of the subsidy per rehoused person that limited the amount local authorities could spend on rehousing and that forced them into the 'minimum standard' housing solution. Further, arguments favouring the retention of higher densities on 'valuable' inner city land, early ideas about the maintenance of working class communities and growing pressures from environmental groups and rural residents to limit urban sprawl, were all given practical expression in the Greenwood Act through the provision of additional subsidy where local authorities chose to build flats rather than houses.

2.5 Twentieth century urban and regional trends and urban renewal

But still state intervention was not the most important influence on urban renewal. The railway and other transport developments continued to exert a dramatic influence over the pattern of urban development. The importance of the Victorian railway building process in changing the physical structure of city centres has already been mentioned, but in many ways a more important and longer term impact of transportation improvements was their effect upon urban densities and the distribution of land values.

Technological changes and increases in the efficiency of transport systems enabled greater distances to be travelled for the same cost, so people and goods could be moved greater distances without incurring cost penalties thus enabling cities to spread over a larger area without economic loss. If a city remained the same size in terms of population and economic activity but covered a larger physical area then the density of development would fall. As densities fell so the average profit from economic activity or rent from housing per unit of area would also fall (i.e. become more thinly spread) and inner area land prices would fall relative to the city average. While the total aggregate value of all the land within the city would remain the same (because the total amount of economic activity and profit had not changed) it would be spread over a larger area. Figure 2.4 illustrates these relationships.

This phenomenon has had important effects upon urban renewal and particularly housing renewal for it has meant that densities have been able to fall (with the benefit of more housing space per person) without loss of aggregate income to landowners as a group.

Until the First World War these transportation improvements mainly took place along the radial routes of tramways and suburban railways, leading to a tendency for cities to follow a star-shaped pattern of expansion. Later however, the impact of the motor vehicle was to accelerate these trends considerably; to fill in the gaps between the points of the star; and to extend rapidly the boundaries of the whole city outwards. The result was a sharp reduction in average urban densities.

The increasing ownership of motor cars permitted suburban and ex-urban housing for the middle classes while the motor bus serviced the growing 'cottage' estates of peripheral council housing; both trends facilitated major reductions in inner city housing densities as housing areas were renewed. The motor lorry permitted the development of the peripheral industrial estates and gave access to large areas of cheap land that enabled

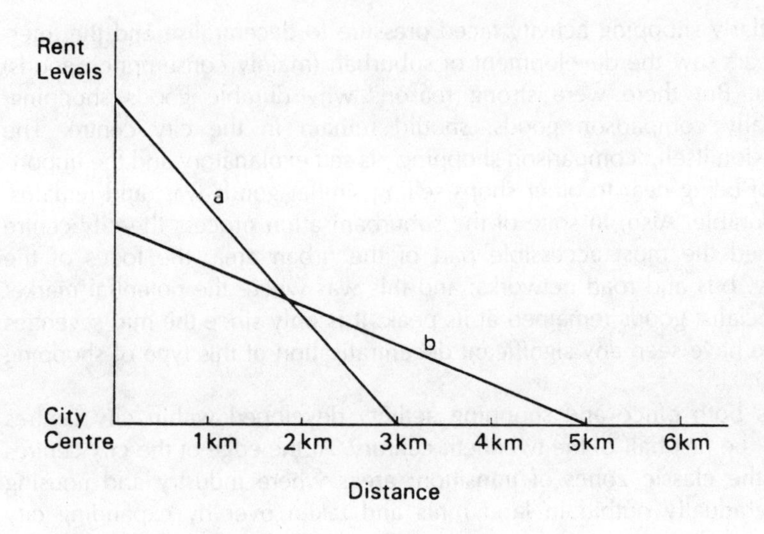

Figure 2.4 The effect of transport improvements on the distribution of land values

new industries to avoid the costs and problems associated with multi-storey buildings within cities. Thus demand for inner city industrial land slackened, enabling it to be used, on renewal, for processes that yielded lower profits per unit of area (i.e. lower densities) and attract relatively lower share of total urban rents.

Only in the city centres themselves were different forces at work. Service sector employment has increased steadily throughout the twentieth century, especially in office jobs. Furthermore, one of the distinguishing characteristics of many office based activities, compared with industrial activities, is the greater need for customer contacts and inter-organisational meetings. Thus while transportation improvements were encouraging office decentralisation, the requirements for personal contact were working in the opposite direction. Into this 'neutral' situation the rapid growth in office employment ensured an absolute expansion of demand for office space in almost all city centres. This growth in demand tended to encourage, on renewal, the intensification of the use of land appropriate for, or (after the introduction of statutory land use planning) zoned for, office use (i.e. to increase densities).

Similarly shopping activity faced pressure to decentralise and the inter-war years saw the development of suburban (mainly consumption goods) centres. But there were strong reasons why durable goods shopping, especially comparison goods, should remain in the city centre. The expression itself, 'comparison shopping', is self explanatory and the importance of being near to other shops selling similar goods was, and remains, considerable. Also, in spite of the suburbanisation process the city centre remained the most accessible part of the urban area: the focus of the railway, bus and road networks; and this was where the potential market for specialist goods remained at its peak. It is only since the mid-seventies that we have seen any significant decentralisation of this type of shopping activity.

Thus both office and shopping activity developed within city centres during the first half of the twentieth century. At the edge of the city centres came the classic zones of transition: areas where industry and housing were gradually outbid in land rents and taken over by expanding city centre activities.

Urban renewal thus became dominated by three main kinds of activity:

(i) state-led housing clearance and replacement building;
(ii) a general reduction in inner city housing and industrial densities which facilitated the takeover of those areas close to the city centre by more profitable office and shopping activities;
(iii) redevelopment of city centres as the density of existing office and shopping areas increased to meet rising demand.

Over the last half-century these renewal processes have taken place within a context of substantial regional shifts in the distribution of employment and population leading to two distinct situations. In some cities urban renewal is taking place in a situation of urban growth while in others renewal is taking place in a situation of falling population and declining demand. The difference between these two situations is important as it presents policy makers and investors in different cities with two quite different sets of economic conditions. In the first case land is in short supply and existing owners of land and buildings are in a strong position to control, manipulate and extract profit from the renewal process; while in the second case it is the land users and developers who are dominant. Figure 2.5 contrasts these two situations.

The regional economic trends that have led to these circumstances are outlined below.

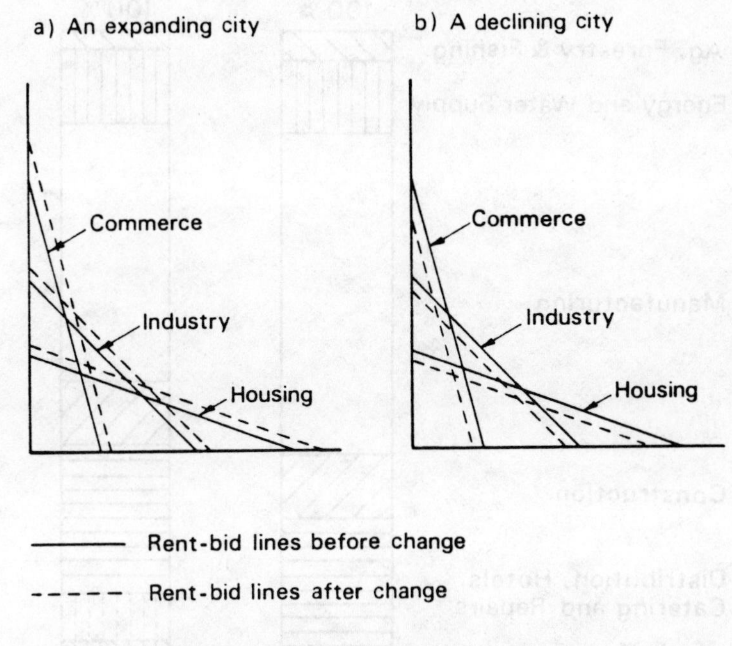

Figure 2.5 The context of urban renewal: city growth and city decline contrasted

2.6 Regional Economic Trends

Recent years have seen a major decline in many of the country's traditional extractive and heavy manufacturing industries. These trends in output are shown in Figure 2.6.

Coal, steel, shipbuilding and railway work, all once vast employers of labour have declined sharply. One feature of this decline has been that these industries tended to be spatially concentrated in the North and North West of England, South Wales, Central Scotland and Northern Ireland. Therefore these regions have tended to be disproportionately affected by this decline.

Secondly the new replacement industries are not generally concerned with the processing or production of bulk materials or heavy goods and are not locationally restricted by access or energy requirements: they are more 'footloose'. In this situation firms have preferred to locate conveniently in relation to their biggest markets, which are seldom concentrated in these regions.

Thirdly almost all industries have achieved substantial increases in productivity in recent years (i.e. more output per worker). In expanding industries this may lead to the same workforce making and selling more goods. For static or declining industries in prosperous areas such as the South East increased productivity is likely to mean that some workers will

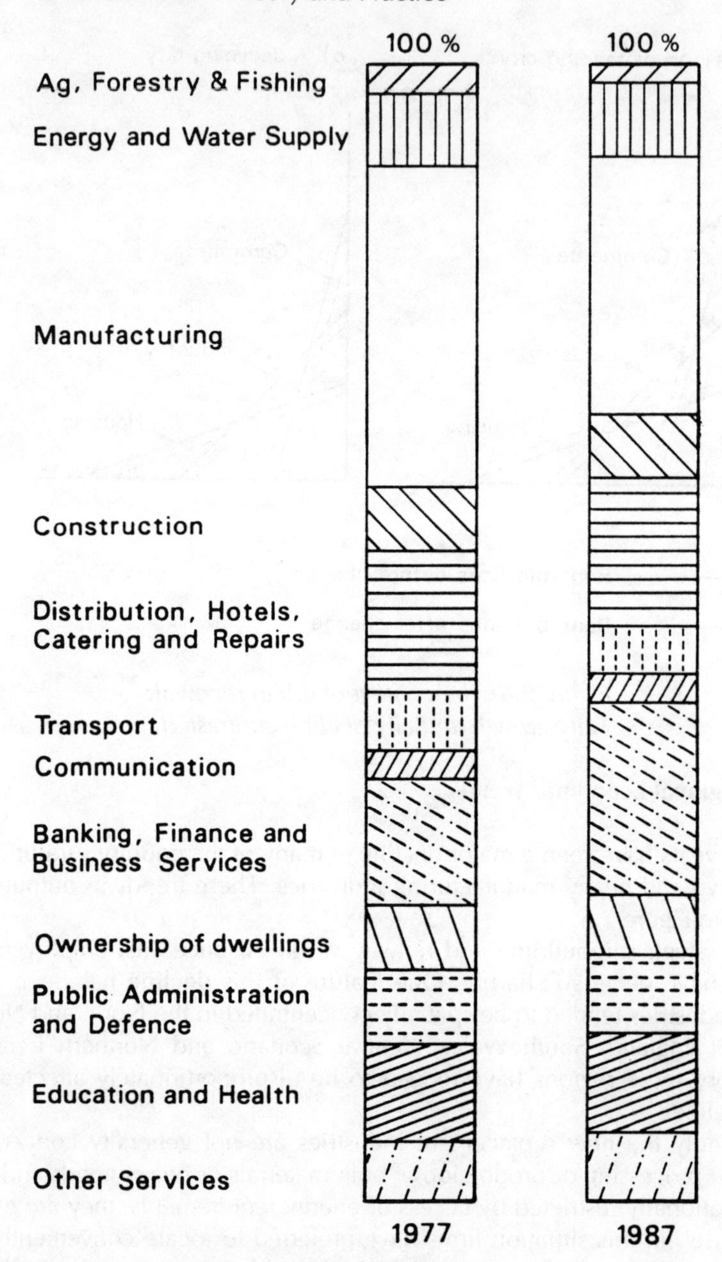

Figure 2.6 Changes in gross domestic product by industry 1977–1987

be laid off but will probably be re-employed elsewhere in the region. In already depressed regions, such as the North or North West, workers laid off will be less likely to find alternative work in the region. Thus productivity gains in steel, coal, shipbuilding and textiles etc. have compounded the already severe effects of falling demand for these products.

Fourthly there have been continuing trends towards industrial concentration and centralisation as less successful firms go out of business or are taken over by more successful firms seeking economies of scale and greater market shares. The result is a general tendency in all industries, for production to become dominated by ever fewer but larger firms. This sometimes, but not necessarily, leads to a spatial concentration of production into fewer locations, but frequently leads to a spatial concentration of senior managerial activity (i.e. power in decision making and the best paid and highest skilled jobs).

Fifthly, there have been major shifts in the location of markets for both intermediate goods and final products. As population has moved from rural areas towards urban areas and from peripheral regions towards more central regions so the market for final sales has moved. These trends have been compounded by the effects of concentration and centralisation, tending to further increase spending power in these growing locations through the general effects of labour shortage and the specific effects of higher proportions of professional and managerial workers. Likewise as industry has grown in these areas so the demand for intermediate goods has moved, creating an incentive for those industries in turn to seek moves or expansion in the same direction.

The state itself has exerted considerable influence on these trends. Through its own expansion of the civil service and related employment; through some aspects of the expansion of defence and scientific establishments; and through the effects of European Community regulations and policies, the state has encouraged the South Eastward drift of population and economic activity. On the other hand for a number of years the government employed regional development policies that attempted to reverse or at least ameliorate these trends. Until recently these policies focussed mainly on attempts to encourage the relocation of firms from prosperous to depressed regions. The policy was at its most successful in the 1960s when the economy was growing at a rate that permitted many firms to expand and open branches or subsidiaries in depressed regions, taking advantage of the various subsidies available. More recently, in times of slower growth, few firms have been in a position to expand and relocate production in this way. Indeed it has been the experience that there has been a tendency for branch-plants and subsidiary firms to close first in times of recession.

For a fuller discussion of regional economic change see, for example, Goddard and Champion (1983).

2.7 Changing Approaches to Urban Renewal Before and After World War II

Looking at the approaches to urban renewal that had developed by the 1950s we find significant advances and changes in thinking from that which dominated in the inter-war period. Some of the influences on this new approach to policy are outlined below.

In 1898 Ebenezer Howard published his now famous book *Tomorrow: A Peaceful Path to Real Reform* (later republished as *Garden Cities of Tomorrow*) in which he argued for a reduction in urban densities through the building of new towns (garden cities) in a 'green belt' beyond the boundary of the central city but still forming part of an economic and social union with that city. In this way Howard envisaged that urban planning could capture the best features (especially the economic and social opportunities) of urban living and couple them with some of the environmental benefits that characterised rural life, while eliminating the worst features of urban deprivation and environmental pollution: hence the term 'garden city'. Howard's ideas were taken up by a generation of planners and others concerned with urban policy, most notably by the Garden Cities Association (which later became the Town and Country Planning Association). This lobby turned out to be one of the most powerful and most succesful in British planning history for it ultimately led to the setting up of the Reith Committee on New Towns which reported in 1946 and the subsequent New Towns Act of the same year. With this legislation the Labour government laid down principles of organisation and management for 'new town development corporations' that have been carried forward today to the Urban Development Corporations now being used to promote inner city regeneration in a number of cities. As Peter Hall has written:

> As for organisation it (the Reith Committee) confirmed that the existing local government structure was not suitable for the task. The new towns, it proposed, should each be built by a special development corporation set up for the purpose, generally responsible to Parliament, but free of detailed interference in its day-to-day management, and with direct Treasury funding.....Almost certainly, Reith's formula was the right one. In building the new towns, freedom for managerial enterprise and energy had to be given priority over the principle of democratic accountability; if the new towns had had to account for every step to a local authority they could never have developed with the speed they did. This was particularly so, since almost by definition the existing community tended to be opposed to the idea of any new town at all.
>
> (Hall, 1975, p.106)

The new towns, together with the later addition of Town Development Schemes and some local authority initiatives such as Liverpool's Kirkby 'New Town', permitted a rapid reduction in inner city working class housing densities and well planned 'garden' living for many as urban renewal accelerated through the 1950s and 1960s.

Just before the publication of the Reith Committee's report but in the same reformist spirit three other reports had been prepared. The Barlow Commission was set up in 1937 (although it did not report until 1940) and was concerned with examining the distribution of population and economic activity across the country; establishing the social and economic costs and benefits of such a distribution and suggesting policies for ameliorating the effects of any 'mal-distribution'. The committee was concerned with the whole issue of urbanisation and urban trends. The report marked a watershed in our understanding of these issues. It quantified for the first time the huge shifts in economic activity that had taken place through the inter-war years and concluded that these movements had been caused by the birth or expansion of firms in particular regions rather than by any significant movement of firms as such. In other words one of the main reasons for growth in the South East was that more firms in expanding sectors of the economy were located there. Barlow recognised that such major movements of population and activity were causing congestion in the South East while leaving underused social and industrial capital in other regions and therefore he felt it would be legitimate for the state to intervene to try and subdue these trends through a regional policy. Such a policy was implemented after the World War II in the form of the 1945 Distribution of Industry Act.

The second of these reports: the Scott Report on *Land Utilization in Rural Areas* was significant, for although mainly concerned with rural matters it expressed grave concern about the rate of loss of countryside under urban sprawl. In order to turn this tide the report recommended that there should be no automatic right to develop and that there should be a comprehensive planning system within which prospective developers had to demonstrate that their proposals were in the public interest.

> Though the onus-of-proof (that the scheme is in the public interest) has never been applied so rigidly in actual postwar planning, there is no doubt that the general sentiment behind the case has been very powerful in supporting the notions of urban containment and of encouraging higher-density urban development so as to save precious rural land.
>
> (Hall, 1975, p.100)

It will be noted that this pressure works in the opposite direction to the garden cities/new towns proposals and was an important influence in the development of high-rise housing renewal solutions, as we shall see later.

The third report, the Uthwatt report, examined the question of compensation and betterment in relation to planning decisions. Betterment, the taxing of unearned profits from land, has been a controversial issue throughout the post-war period with three separate attempts to implement a policy (under the 1947 Town & Country Planning Act, the 1967 Land Commission Act, and the 1975 Community Land Act) all frustrated and repealed by subsequent governments.

During and immediately after World War II plans were prepared for the reconstruction and redevelopment of many cities. Some of these plans were remarkable for their perception of the problems and for the boldness of their solutions.

In London various self-appointed groups were set up to prepare such plans: the RIBA sponsored London Regional Reconstruction Committee; the Royal Academy; and the Modern Architectural Research Group all made proposals. In addition two official plans were prepared for the planning and reconstruction of the capital: both the 1943 County of London Plan and the 1945 Greater London Plan were prepared by teams led by Sir Patrick Abercrombie, one of the most influential and respected planners of the time.

The basic approach adopted by Abercrombie in his 1945 plan was to divide the conurbation into a series of concentric rings. In the inner ring, most of the pre-1919 city, urban renewal would lead to a reduction in densities and an 'overspill' of displaced population and economic activity in new towns beyond a 'green belt'(a ring of preserved countryside) around the edge of the existing built up area.

The approach to the issue of urban renewal being adopted by most planners at the time had three important characteristics:

(i) There was a strong assumption that the state would either undertake or have a powerful control over most redevelopment.
(ii) There was a high level of confidence that planners 'knew what they were doing' and that their technical-physical approaches provided valid solutions to urban renewal questions.
(iii) There was a far greater emphasis upon urban design and aesthetic values than there had ever previously been in town planning reports and a concomitant lack of concern, or naivety, about social and economic matters.

These features are to be found in many plans of the era. Typical of the genre is the 1947 Outline Plan for the County Borough of Birkenhead prepared by Professor Sir Charles Reilly and N J Aslan: a bound 200 page volume with colour printing and glossy photographs, it exudes the confidence of the era. Urban renewal is considered in section 7 under the title 'The Replanning of Certain Areas' and in section 10 'Proposals for Brightening the Town'. In the first of these sections there are policies for the rebuilding of vast areas of the inner city (most of which was *not* war damaged). These proposals include the establishment of great set pieces of urban design: squares; focal points; vistas; avenues and parks; with the housing and other buildings to be fitted into this rigid physical framework. In section 10 amongst the proposals is the preservation of the character of areas:

> we are suggesting that ... instead of pulling down the big houses there [Bidston Hill] each with several staircases and three or four acres of garden, and allowing little houses to be built in their stead, the character of the neighbourhood be preserved by turning the gardens into a public park and the big houses into maisonettes
>
> (Reilly and Aslan 1947, p.117)

There is no mention of how this is to be carried out, what the legislative powers or sources of funds would be, but the proposals are intriguing for their imagination. The plan also proposed smoke control: it is easy to forget the impact that domestic coal fires and commercial steam power had on the environment of towns before the Clean Air Acts, natural gas and electricity dominated cities. Other policies were put forward for control of advertising and the exterior painting of buildings and there was a list of places where special architectural (civic design) opportunities were felt to exist. One such example, the 'Woodside Hotel' site is illustrated in Figure 2.7 together with the original caption. Figure 2.8 summarises the development of thinking and influences on post-war urban renewal thought and policies.

2.8 Post-war Slum Clearance and High-rise Housing

After the initial post-war reconstruction period, as the economy improved and the shortage of construction materials eased, attention turned again to the business of relieving overcrowding and poor housing conditions in the major cities. Closely connected with the subsequent period of rapid slum clearance and housebuilding is the story of the nation's disastrous embrace of high-rise housing and industrialised building techniques: a story that reveals much about decision-making processes in housing and urban renewal policy at that time.

Figure 2.9 shows the importance of high-rise housing in the post-war period: its rapid acceptance into mainstream housing provision and its later equally rapid abandonment.

The 1930s slum clearance movement created a climate of acceptance, at least by the state, of flats and tenements as a suitable building form for replacement working class housing. In spite of opposition from some expert opinion local authorities, particularly in the big cities, rapidly

Figure 2.7 1947 proposals for the Woodside Hotel site, Birkenhead. The contemporary caption read: 'It abuts a main line railway terminus which could well become, with the construction of a Channel Tunnel, the western end of the European railway system.'

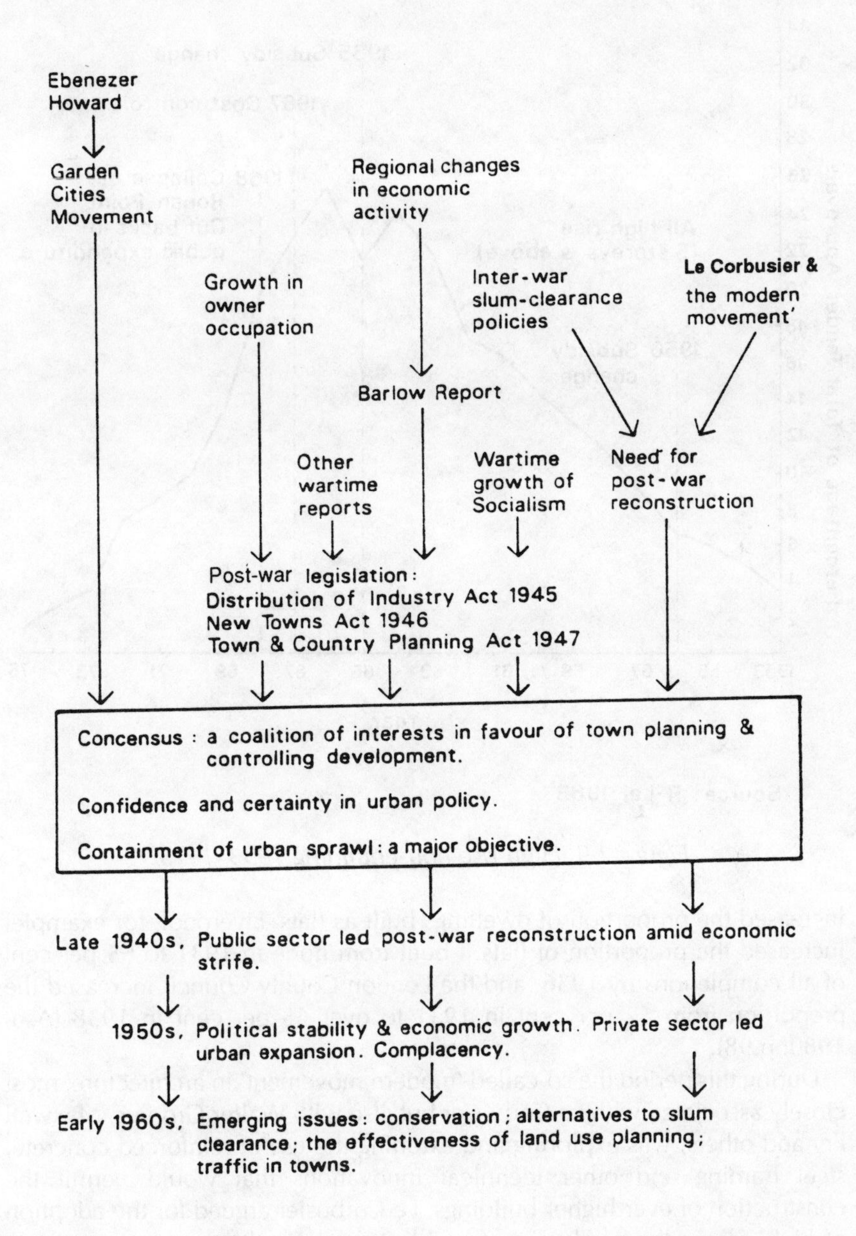

Figure 2.8 Influences on post-war urban renewal thinking and policies

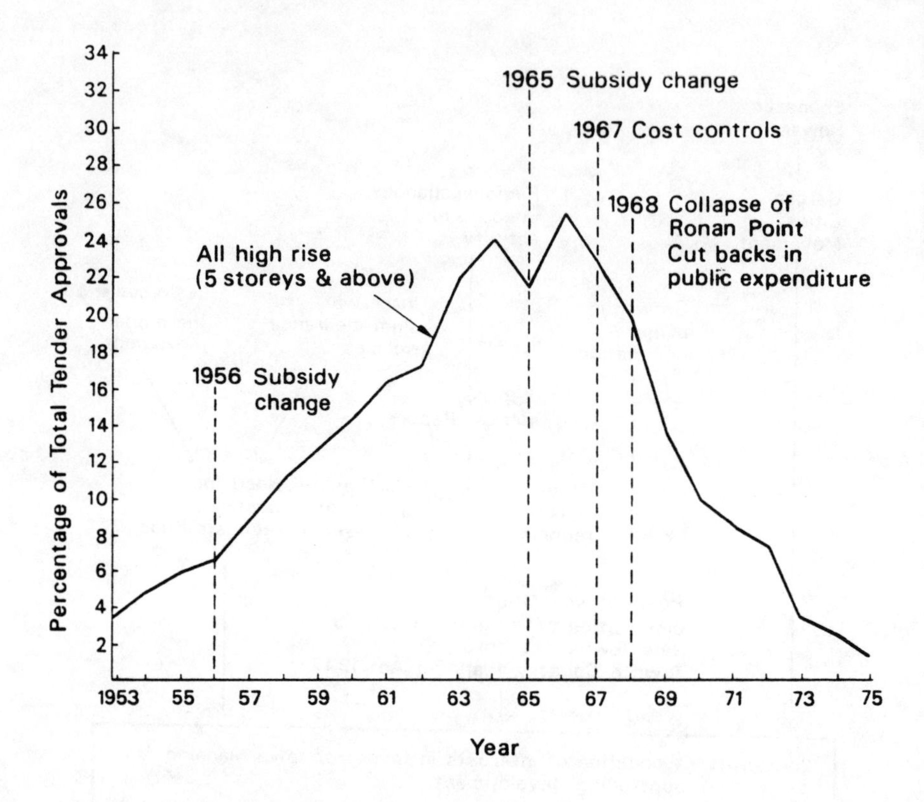

Figure 2.9 High-rise housebuilding 1953–1975

increased the proportion of dwellings built as flats. Liverpool, for example, increased the proportion of flats it built from none in 1931 to 65 per cent of all completions by 1936, and the London County Council increased the proportion from 12 per cent in 1931 to over 65 per cent in 1938 (Ash, 1980, p.98).

During this period the so-called 'modern movement' in architecture, most closely associated with Le Corbusier but also with Walter Gropius, Maxwell Fry and others, was exploring and extolling the use of reinforced concrete, steel framing and other technical innovations that would permit the construction of ever higher buildings. Le Corbusier argued for the adoption of high rise as the predominant building form for the modern city while Gropius and others envisaged architects working much more closely with the building components industries to produce 'industrialised' building forms that could be manufactured under factory conditions of precision and efficiency and assembled on sites. This movement had a profound and pervasive effect upon the architectural profession, especially, as Cooney

(1974) points out, on those who studied in this period and who were occupying senior and influential positions in the post-war years.

By the early 1950s there was general acknowledgement of the severity of the housing shortage facing the country and both the rapid expansion of the housing stock and the renewal of its worn out parts were high on political agendas.

In 1951 the London County Council completed the 'Roehampton Estate', a mixed development of high- and low-rise accommodation in a picturesque parkland setting in south-west London. This development gained a world-wide reputation and won many design accolades. It demonstrated that high-rise, high-density local authority housing could be designed and built to provide good living conditions and helped to legitimate the design

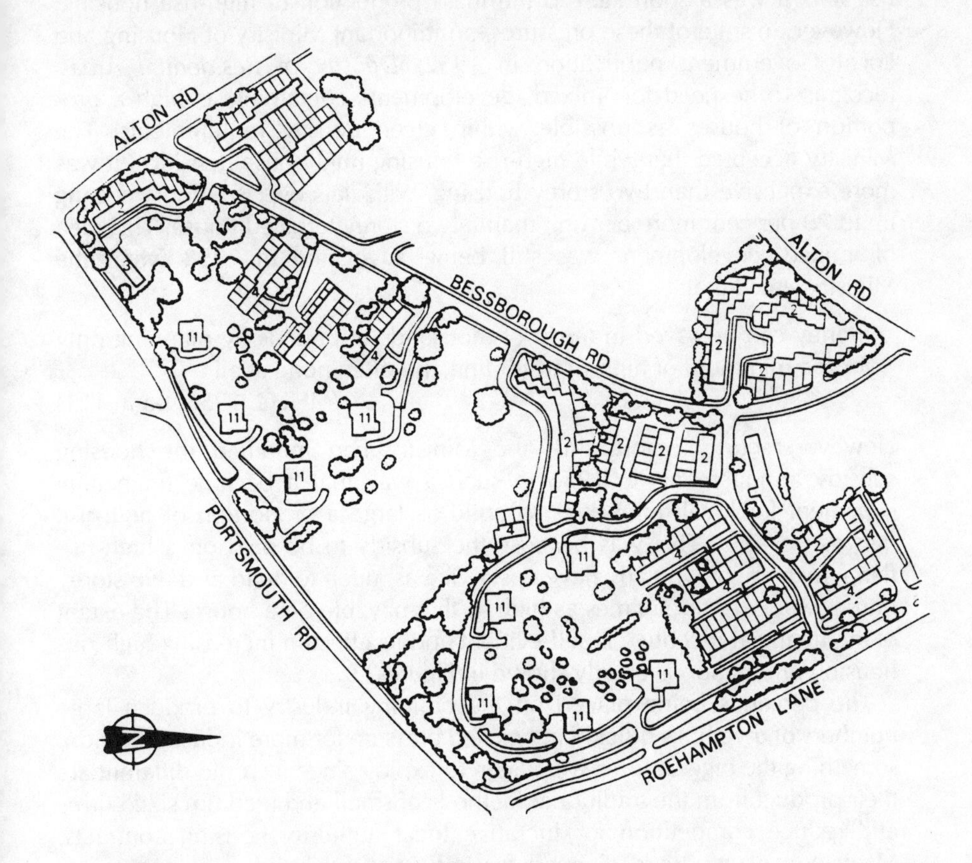

SITE PLAN. The numbers indicate storeys.

Figure 2.10 The London County Council's Roehampton Estate

philosophy of the modern movement. The scheme is illustrated in Figure 2.10.

One of the stronger aims of town planning at this time was the control of the adverse visual and economic effects of urban sprawl. There was also a strong strategic concern to improve the degree of national self-sufficiency in food production and therefore to retain as much land as possible in agricultural use. Added to this was a reluctance by many rural districts in the vicinity of major cities to accept local authority overspill estates and an equal reluctance by some cities to lose population for fear of the effect upon the rate support grant and service provision. Thus there was widespread support for the idea of urban containment and raising the densities of slum clearance and overspill estates.

Given the desire to build at high densities and the legitimation of high-rise flats it was a short step to the mass production of high-rise housing. However, in spite of these pressures, an important Ministry of Housing and Local Government publication in 1952, *Density of Residential Areas*, recognised the need for 'mixed' developments containing as high a proportion of houses as possible within given density requirements. The Ministry accepted that while high-rise housing might raise densities, it was more expensive than two storey housing, with flats over six floors costing up to 70 per cent more per unit than a two storey terraced house. This idea of 'mixed' development was still being advocated in 1958 when the Ministry stated that:

> money can be saved in terms of thousands of pounds by planning only for the minimum of high-rise building and often none at all
>
> (MHLG, 1958, para. 121)

However, shortly before this, the Ministry had modified the housing subsidy arrangements in 1956 in such a way that it became financially expedient for local authorities to build as large a proportion of high-rise flats as possible. This was because the subsidy to be paid on a high-rise flat, which might typically have cost twice as much to build as a two storey house, was over three times as high as that payable on a house. The extent of the financial incentive is self-evident and its effect in increasing high-rise housing production is clearly shown in Figure 2.9.

The demands being placed on the building industry to produce large numbers of dwellings quickly intensified pressure for more industrialisation, something the biggest firms were keen to exploit since it would differentiate their product from the traditional methods of small and medium sized firms and reduce competition for lucrative local authority housing contracts. Many advantages were claimed for industrialised building including reduced production times, improved productivity and lower costs. However, industrialised mass production also required large and assured markets and limited the design opportunities available since the nature and size of most components was pre-determined. These pressures caused changes in the relationships between local authorities and the construction industry. Large firms with a patented industrialised building system offered a 'design and build' or 'package deal' contract for hundreds or thousands of dwellings at a time. In consequence contract size increased, the number of competing contractors reduced and local authority architects were increasingly excluded from the design process. Once a contract had been signed the power of the contractor increased sharply while that of the client local authority diminished.

The final consumer of local authority housing, the occupier, had always had a weak voice in this process but the effect of these new contracts was to render even their representatives, councillors and professional officers, ineffective in controlling the quality of design and construction. Nevertheless, as people began to live in these blocks and as housing authorities began to appreciate the true costs of running such flats there began a learning process that was very quickly to lead to an end to high-rise and industrialised building in this country.

It had been known that flats would initially be more expensive to build but the major contractors had held out the promise that scale economies and standardisation would bring down unit costs in future years. By 1965 the future still seemed a long way off and it was becoming clear that normal maintenance and management costs were much higher than for houses and that high-rise blocks were beginning to experience a large number of voids and an alarming number of problems with rain penetration, condensation, broken lifts, blocked rubbish chutes and other serious defects. Many of these problems arose from the fact that the assembly of system components often required on-site labour to work in ways and to a degree of accuracy largely unknown and unnecessary in traditional construction.

By this time there was also a growing view that flats provided unsuitable accommodation for raising young children in. Gradually the weight of evidence forced local authorities to stop allocating high-rise accommodation to further families and to begin moving existing families to more appropriate accommodation.

Another argument that proved false was the idea that high-rise building saved land. The publication of Ministry advice in *Residential Areas: Higher Densities* (MHLG, 1962) showed the fallacy of the argument, although similar information produced by Lewis Keeble was published and available as early as 1952 (Keeble, 1952, Chapter 17)! However, the events that finally spelt the end of the high-rise movement were the introduction of unit cost controls by the Ministry in April 1967 and the balance of payments crisis later the same year. The effect of the housing cost yardstick was to make high-rise developments uneconomic since it became almost impossible to build them within the prescribed cost maxima. The nation's economic difficulties led to major cutbacks in public expenditure including over £100 million off the building programme. Public sector housebuilding was reduced from 170,000 dwelling approvals in 1967 to only 112,000 approvals in 1969. The effect of this cutback was to shift the balance of economic power away from the major builders back towards the local authorities. Contract sizes became smaller and there was some reversion to more traditional building methods. Both moves brought small and medium sized builders back into local authority housebuilding and returned more design and contract supervision to local authority control.

The most dramatic event of the high-rise saga was the collapse of Ronan Point, a twenty-one storey system-built block in the London Borough of Newham, in May 1968. A gas explosion in one flat led to the progressive collapse of one entire wing of the block, killing five people. The Tribunal of Inquiry set up to investigate the tragedy showed up a whole range of disturbing faults in the operation of the industrialised building drive; but long before its publication public outcry and political expediency had spelt the end of high-rise housebuilding.

2.9 The Move from Clearance to Improvement

The end of the sixties marked another significant turning point in housing renewal as policy began to shift away from large scale slum clearance programmes towards rehabilitation and area improvement.

In the wake of the 1967 economic crisis and the downturn in public sector housebuilding it was becoming obvious that the slum replacement building programme was not keeping pace with the ambitious rate of clearance set by many authorities, so increasing pressure on the existing stock and generating ever growing zones of vacant land awaiting redevelopment.

There was also growing dissatisfaction with slum clearance as a method of dealing with obsolete housing. A number of issues were identified:

(i) Until the 1960s, as a broad generalisation, clearance programmes dealt with mid-Victorian housing of very poor quality. After this time many cities were beginning to include superior quality housing from the late Victorian period within clearance areas. Such moves proved controversial since it was argued by professionals and residents that these houses were often structurally sound and with some renovation could provide adequate, even good, housing. A number of official studies (e.g. at Deeplish in Rochdale (MHLG, 1966)) and academic studies (e.g. Needleman, 1965) attempted to show the feasibility and economic benefits of renovation.

(ii) A growing conservation movement was becoming active on many fronts awakening public opinion to problems of limited energy resources, the destruction of increasingly vulnerable natural habitats and landscape, and the importance of civic and cultural heritage, including the built form and townscape of urban areas. With this last concern came an appreciation that even the Victorian city, which until then had generally been assumed to have almost no features worth preserving, should be more sensitively conserved and subject to only piecemeal redevelopment.

(iii) The clearance programme itself had disappointed many people. What had been heralded as 'a new beginning for cities' and a promise to provide all families with a decent home came in for two particular criticisms. Firstly, the way many local authorities dealt with the inhabitants of clearance areas was often regarded as high-handed and insensitive. Given the scale of some programmes, perhaps rehousing as many as ten families a day, and given the limited calibre and training of housing department staffs at that time, this was not suprising but nevertheless an unacceptable state of affairs. Secondly, there was disillusionment with the nature and quality of the replacement dwellings: high-rise flats, remote locations, excessive fuel bills and poor construction seemed an inadequate substitute for the insanitary but convenient homes many families had left behind. Of course not all redevelopment programmes were like this, many thousands of people were pleased with their moves, but there was a sizeable proportion who were dissatisfied with their changed circumstances, especially as the initial excitement of moving wore off.

(iv) There was also a growing feeling that many of the areas earmarked for clearance housed communities that were longstanding, close-knit and contained features of social life that were worth preserving. While the sociological evidence, both then and subsequently, suggested that this was a gross simplification, it was a sufficiently powerful argument to influence opinion at the time.

A further social trend that was influential in the strategic thinking of big cities was that of population decline. By the time of the publication of the 1971 Census it was becoming clear that most British cities were heading for a future of substantial population decline. This caused local councils to think hard about clearance and especially the inevitable associated overspill and loss of population to neighbouring authorities. In this climate an alternative policy that would retain population, rateable values and rate support grant, was an attractive proposition.

By the time the major slum clearance programmes had come to an end the whole fabric of many inner city areas had been transformed from the close-knit dense pattern of Victorian streets with their serried ranks of two-up two-down terraced housing, corner shops and pubs intertwined with the occasional back street factories, into single tenure 'planned' neighbourhoods of high- and mid-rise flats and windy shopping precincts in which the concept of the street had all but vanished. Figure 2.11 shows the transformation of Everton in Liverpool caused by the slum clearance and rebuilding programme. It is worth noting that many of the tower and slab blocks of flats shown on the second plan have been demolished by the City Council over the last five years in an attempt to re-humanise the area.

2.10 Housing Improvement Policy

Housing improvement has been an element of urban renewal policy for forty years since the 1949 Housing Act introduced Discretionary Grants of up to 50 per cent of the total cost (up to certain maxima) for the refurbishment and improvement of dwellings. In 1959 this system was supplemented by Standard Grants payable under the House Purchase and Housing Act 1959 for the provision of essential basic amenities (water

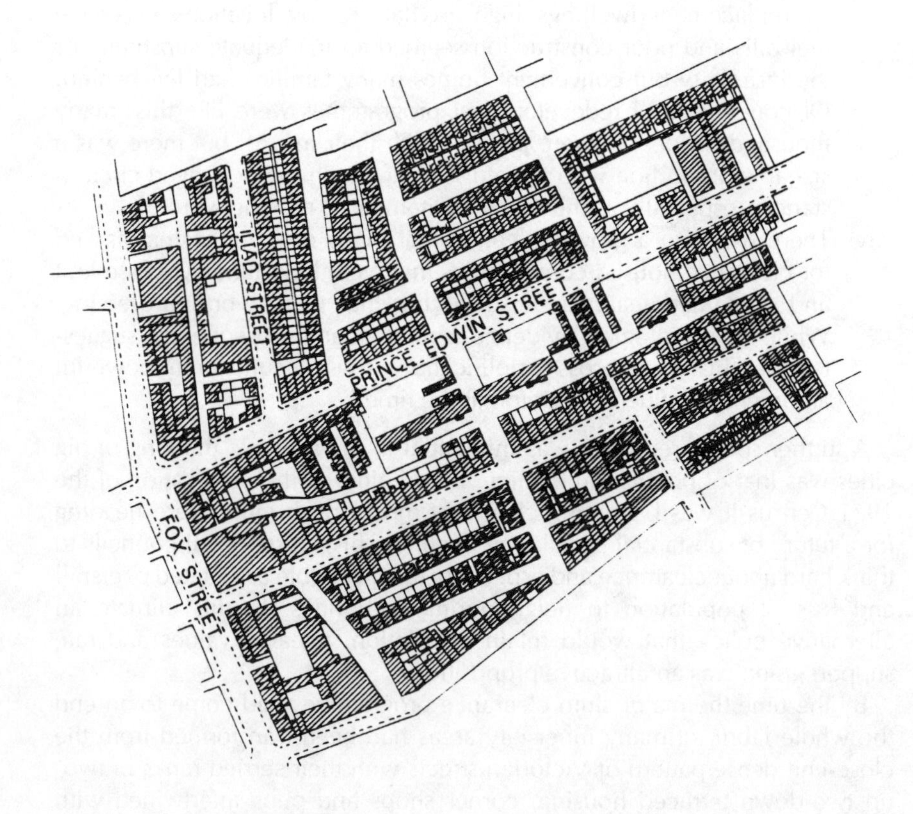

Figure 2.11 (a) Everton, Liverpool, before slum clearance and rebuilding

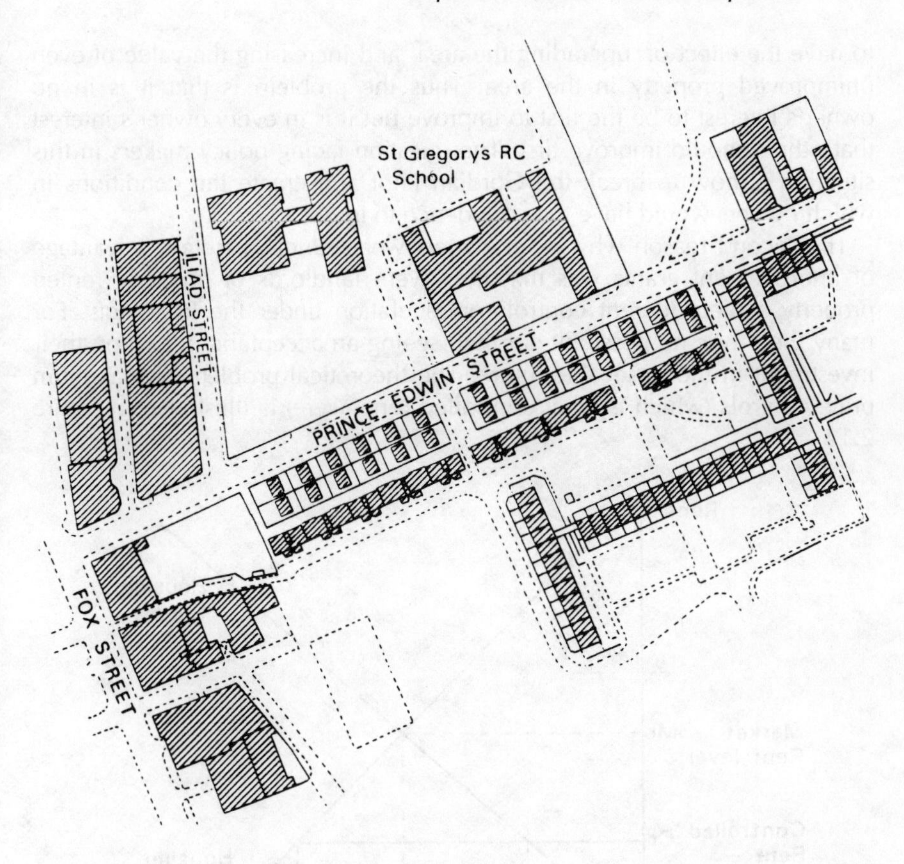

Figure 2.11 (b) Everton, Liverpool, after slum clearance and rebuilding

supply, WC, etc.) in dwellings where they were lacking. Together these grants accounted for a steady rate of around 120,000 to 130,000 dwelling improvements a year throughout the sixties. One of the perceived shortcomings of improvement policy was that owners, especially landlords of rented properties in slum areas, were reluctant to invest while uncertainty existed about the future value of their investment.

There were two problems. Firstly, all owners were faced with two sorts of uncertainty: (i) would their property or adjoining property be included in a clearance scheme at some future time and (ii) could they be sure that neighbouring owners would also invest in property improvements? This second problem is known as the prisoner's dilemma. The basis of the argument is that if a houseowner carries out improvements to his or her dwelling and the neighbours don't, then even though the owner's property has been improved the increase in value will be held back by the general deterioration of the area and may not justify the investment. On the other hand if many neighbouring owners carry out improvements, they are likely to have the effect of 'upgrading the area' and increasing the value of even unimproved property in the area. Thus the problem is that it is in no owner's interest to be the first to improve but it is in every owner's interest that others should improve first. The question facing policy makers in this situation is how to break the Gordian knot and create the conditions in which owners would have the confidence to invest.

The second reason why some owners were reluctant to take advantage of improvement grants was that they were landlords of privately rented property subject to rent controls or regulation under the Rent Acts. For many there was no possibility of their seeing an acceptable return on their investment in the forseeable future. The theoretical problem of maximum price controls (which is what controlled rents were) is illustrated in Figure 2.12.

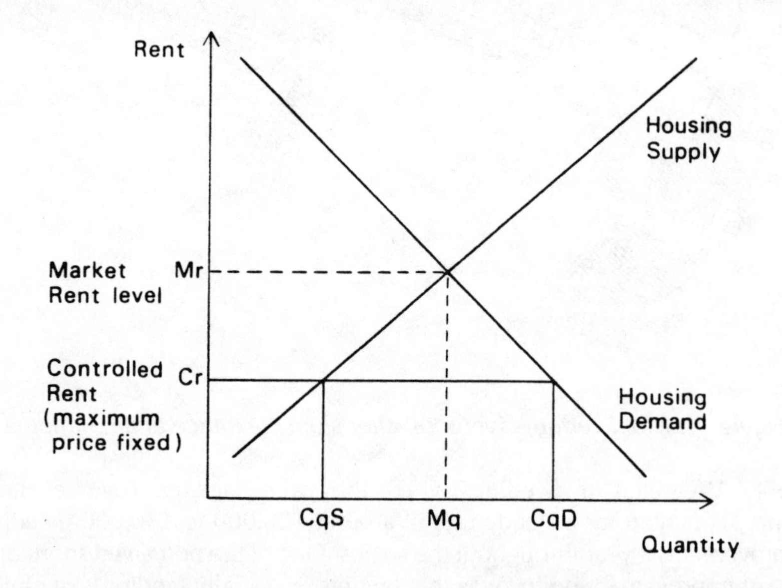

Figure 2.12 Maximum prices and rent controls

The effects on housing supply in this sector are self-evident. Few new landlords entered the sector and those already supplying private rented housing were generally reluctant to spend money on normal maintenance and minor repairs, let alone major improvements. The problem of the distorting effects of rent controls and the question of how to bring about improvements in the private rented stock was one that remained peculiarly unresolved until the most recent round of housing legislation, and even here the solution may prove illusory.

One response to the second problem was to introduce the possibility of compulsory improvement of rented property under the 1964 Housing Act although, for a variety of reasons, little real progress was made in this direction. More positive moves were made to solve the first problem. Under the Housing Act 1964 local authorities could declare 'Improvement Areas' in order to remove the threat of clearance from suitable areas, create a stable investment climate and encourage 'comprehensive improvement'.

There were two further reasons why the government was keen to encourage 'area improvement'. One was the perception that bad housing conditions were spatially concentrated and so the targeting of resources would have a greater proportional effect than distributing the grants randomly across the older housing stock. This argument, which was generally accepted at the time has weakened in force in recent years as precisely these area initiatives have improved large concentrations of housing but left isolated pockets and individual unimproved dwellings scattered across urban and rural areas. It was also felt that there were economies of scale to be gained in concentrating resources in specific areas. While it was the case that local authorities did let large contracts for the uniform improvement of whole streets or blocks of council owned dwellings and must have achieved some scale economies, most private sector improvements were carried out on a one-off basis by individual owners and in cost terms, benefitted little, if at all, from the fact that other neighbouring owners were also improving. Indeed it is more likely that the concentration of demand for scarce specialist building skills within a confined area had the opposite effect of pushing up building costs.

With the passing of the 1969 Housing Act the government refined area improvement policy through the introduction of 'General Improvement Areas' (GIAs). Within designated GIAs, in addition to existing powers, local authorities were obliged to consult local residents about improvement proposals and were given an allowance of up to £100 per dwelling to spend on environmental works (landscaping, parking provision, streetworks, etc.). (A typical example of General Improvement Area proposals can be seen later in Figure 6.1) Despite this the rate of grant take up increased only slightly on previous levels until the passing of the 1971 Housing Act which increased the level of grant payable from 50 per cent to 75 per cent of eligible expenditure in all Assisted Areas (Intermediate Areas, Development Areas etc.). This higher level of grant was to be available for three years until 1974. The effects were dramatic. By halving the amount of money property owners had to find for improvements the rate of improvements was more than doubled within a year. Figure 2.13 shows the numbers of renovation grants approved from 1963 until 1987. The effects of the 1971 legislation can be clearly seen.

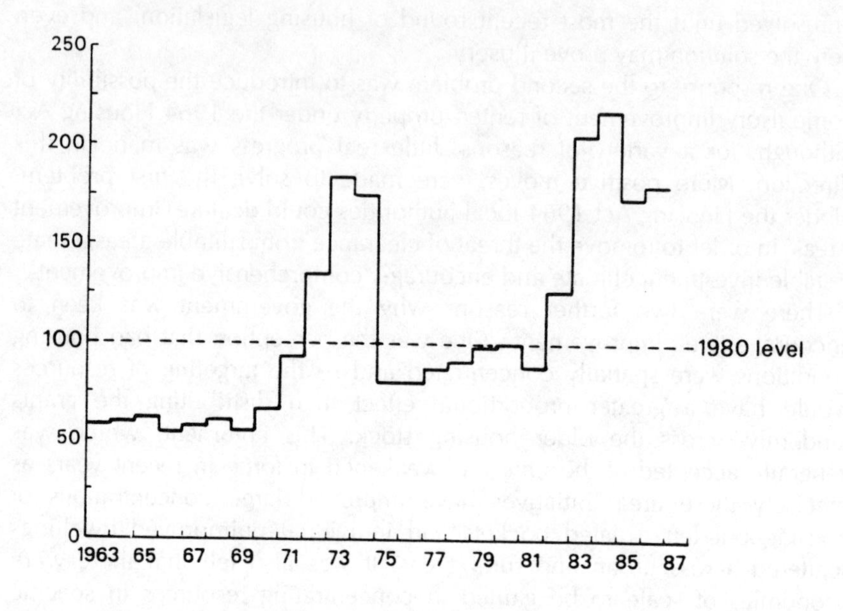

Figure 2.13 Housing renovation 1963–1988

The main purpose of the 1971 grant changes was to stimulate the building industry; they were economic measures with economic rather than housing objectives. Thus while they were demonstrably successful in stimulating building activity they were criticised by many on the housing side for undermining the idea of concentrating improvement activity in the worst areas and allowing scarce resources to be 'squandered' on improvements to better quality suburban and rural housing and even second homes and holiday cottages.

By the time of the preparation of the 1973 Housing White Paper it was possible to review the performance of housing improvement policy and make a number of general comments. Most grant take up was outside designated General Improvement Areas. Many grants were taken up in the owner occupied sector and the local authority sector but few in the private rented sector. Many grants were taken up for the improvement of the inter-war stock while the take up for use on older property was disappointingly low. There was also some evidence of abuse of grants where they were being used for improvements to second homes and in the gentrification of property for profit, mainly in inner London. The general view of GIAs was that while there were notable examples of good practice and individual success there was not sufficient scale of activity to have more than a marginal effect on the general condition of the housing stock within most urban areas.

By 1974 a degree of bi-partisan agreement had been reached between the main political parties on the matter of housing renewal policy. So much so that many of the ideas contained within the 1973 Conservative White Paper 'Widening the Choice: the Next Steps in Housing' were carried forward into legislation by the incoming Labour administration. Prominant amongst these proposals was the concept of Housing Action Areas. This was an attempt to overcome two of the biggest criticisms of existing housing improvement policy: to focus attention and spending on those areas where the need for housing improvement was greatest, and to encourage greater levels of improvement in the private rented stock. Local authorities could designate pockets of housing *and* social stress as Housing Action Areas where a five year 'crash programme' of housing improvements would be instigated using grants of up to 75 per cent of the cost of improvements (rising to 90 per cent in cases of hardship). This was explicitly a 'worst first' approach to housing renewal and extended the area based approach to embrace social stress: a translation of deprivation area and cycle of deprivation theories into policy.

The first effect of the 1974 changes was to reduce drastically the amount of housing improvement as the higher levels of grant were withdrawn from suburban and rural areas. It took some time for the urban authorities to designate Housing Action Areas and to gear themselves up to the intensive and subtle management that the programme required in these difficult inner city areas. In some ways the verdict on HAAs has been similar to that on GIAs. In spite of much local success in the implementation of individual HAAs, the level of resources devoted to the problem by central government was insufficient and too erratic to do any more than stop average levels of disrepair from getting any worse.

2.11 The Inner City Question

As we have already seen, by the mid-sixties there had arisen a debate about the effectiveness of slum clearance and arguments had been put forward favouring the retention and improvement of existing older residential areas. Indeed the whole question of inner city living conditions had risen to the top of the political agenda (the Milner-Holland Report on Housing in Greater London was published in 1965, Shelter was founded in 1966 and media interest was rising, for example the famous *Cathy Come Home* television programme in 1967 movingly illustrated the plight of the homeless). One consequence of this was that inner city social and economic problems could no longer be bulldozed away and dispersed elsewhere but would need to be tackled *in situ*.

At the same time growing racial tension in the inner cities, resulting from local reaction to waves of new Commonwealth immigration since the early 1950s, forced the government (initially through the Home Office, since the problem was perceived as one of social disorder and control) to establish a programme of Urban Aid. Under this programme local authorities and community organisations could bid for both capital and revenue funds to support social and community development projects. In the same year (1968) the Ministry of Education designated a series of Educational Priority Areas in selected inner cities, within which additional funds were aimed at improving educational provision and attainment. The logic of the approach was that children in these areas, which were based upon school catchment areas, performed worse than average at school because of a cycle of deprivation and disadvantage in which their poorly educated, low income and inadequately housed parents placed a low value on education and offered little extra-curricula support. By breaking this cycle through the education system it was argued that such children might achieve better jobs with higher incomes and escape from the inevitability of continuing poverty (Lawless, 1979, Ch.3).

In 1969 the Home Office publicly launched an investigative study into the nature and causes of deprivation through the Community Development Project (CDP). What started as a series of isolated local action–research community development projects turned into a major social research movement. It went far beyond acceptance of the current conventional wisdom of 'culture of poverty', 'cycle of deprivation' or 'managerial' explanations for urban deprivation to produce strongly argued reports clearly locating the causes of relative poverty in fundamental changes in the economic structure of cities and in the imperatives of the capitalist economic system. Any solutions that did not address these fundamental questions were seen as merely ameliorative and likely to be of little long term value. (See: Loney, 1983 and various National CDP publications such as *What Ever Happened to Council Housing?*, *Costs of Industrial Change* and *Gilding the Ghetto*.)

The Community Development Project lasted nearly ten years with most of its influential and more radical publications emerging in the mid-seventies. By this time two other strands of investigation had begun to reinforce their analysis of the inner city if not their prescription. Firstly, in 1972 the Department of the Environment appointed consultants for three major Inner Area Studies (IAS) in Liverpool, Lambeth and Birmingham, to examine the general nature of inner area problems and to recommend policies. Their reports began to appear in draft form from 1975 with final reports being published in 1977 (DOE, 1977, a, b, c). Secondly, under the 1968 Town & Country Planning Act, Structure Plans were being prepared for the new Metropolitan Counties. One requirement of this plan preparation process was the investigation of social and economic problems within Structure Plan areas. Each of these metropolitan areas contained severely deprived inner cities which became a focus for the planners' attention. While Structure Plans were essentially land use planning mechanisms with little power to tackle the social and economic issues, many of their accompanying reports of survey contained a wealth of information and analysis which tended to reinforce the conclusions of the CDPs and IAS as to the fundamentally economic nature of the inner city problem, even though there were differences in policy recommendations (considerable in the case of the CDPs).

The so-called *Stage One Report* of the Merseyside Structure Plan illustrates the problems facing the planners:

> As an older urban area which has experienced an economic decline relative to the national economy, the County is not faced with growth but with static or declining levels of activity. In this situation, the basic planning powers of development control and land allocation, designed to control new development where the market pressure was assured, are less important than policies which initiate redevelopment and regeneration.
>
> (Merseyside County Council, 1975, p.1)

The government response came in 1976 and 1977. In a famous speech in Manchester on 17 September 1976 Peter Shore, then Secretary of State for the Environment, acknowledged the nature and growing severity of the inner city problem:

> All our major cities have lost population over the last decade and a half. Since 1961 the inner area of the Manchester conurbation has lost 20 per cent of its population and that of Liverpool 40 per cent. What is more worrying is the unbalanced nature of the migration .... leaving the inner areas with a disproportionate share of unskilled and semi-skilled workers, of unemployment, of one-parent families, of concentrations of immigrant communities and overcrowded and inadequate housing. Though there has been a growth of office jobs in the centres of most of these cities this has not compensated for the extremely rapid decline in manufacturing industry in the inner urban areas. Manufacturing employment in Manchester declined by 20 per cent and Liverpool by 19 per cent between 1966 and 1971.
>
> (Shore, 1976)

Shore's response was impressive: the needs element of the rate support grant was adjusted to give increased assistance to inner city local authorities; the new and expanded towns programme was cut back so as to reduce out-migration from the cities and the 1978 Inner Urban Areas Act placed inner city policy in the mainstream of central government concern. Under this change in policy Urban Aid was relaunched as the Urban Programme with increased funds and transferred from the Home Office to the Department of the Environment. A series of 'partnerships' were established under which central and local government agencies would work together to manage a coordinated programme of investment in all the most deprived inner cities. A second tier of less severely affected 'programme' authorities was also designated. The Act also permitted local authorities to designate Industrial and Commercial Improvement Areas.

Even before the 1979 election questions were being raised about the effectiveness of these partnerships in times of recession and public expenditure cuts and the relevance of this 'managerialist' (i.e. changing government organisational and service delivery arrangements) solution to problems that many had analysed as requiring more fundamental economic solutions. But the arrival of the new Conservative government with its tough supply side economic policies, its distrust of Labour controlled local authorities and its different priorities, changed the whole emphasis of inner city policy.

In 1980 under the Local Government, Planning and Land Act the government set up Urban Development Corporations (UDCs): an application of the new towns development corporation organisational model to the urban regeneration situation. Initially two UDCs were created: one in the London Docklands and one in Liverpool. These development agencies were given very clear tasks in relation to removal of physical dereliction, bringing land back into beneficial use and facilitating property development. Neither job creation nor the provision of housing or social facilities to meet local needs were seen as primary objectives. This property led approach to urban regeneration was later extended to other cities as UDCs were established in the West Midlands, Cardiff Bay, Teesside and most other conurbations.

Another wing of this attack on reducing developers' costs was the idea of Enterprise Zones announced in the 1980 budget. Within designated zones occupiers and developers benefitted from certain tax concessions and a relaxed planning regime. The early eighties also saw the government attempt to limit local government spending and local rate increases, partly for macro economic reasons and partly in a belief that high inner city rates were deterring regeneration.

In 1981 localised inner city riots in several cities temporarily shook the confidence of the government and led to a high profile response which included Micheal Heseltine, then Secretary of State for the Environment, for a short period being styled the 'Minister for Merseyside' and the setting up of the Merseyside Task Force, a local inter-ministerial team to coordinate central government activity; and a series of initiatives designed to get more private sector investment into the inner cities. This latter development has come to be one of the most important planks of inner city policy: encouraging, cajoling and subsidising private sector investment. The results of this have been a change in the nature of 'partnership' from being a central–local government relationship to being a public–private sector relationship and a burgeoning of profit seeking development companies keen to exploit this new found investment market. The consequences for the inner cities have been a sharp increase in *ad hoc,* sometimes uncoordinated and unplanned property development activity, and a number of significant improvements in the local physical environment, but frequently with scant regard to the social and employment needs of the indigenous local community.

2.12 Conclusions

Even from this brief overview of the evolution of urban renewal in Britain it is clear that there are two distinct but interlinked processes going on. The first is that urban areas are constantly being reconstructed in order to maintain the efficiency of the capitalist system, including the maintenance of adequate markets for goods and services. In this process there is always pressure from capital to improve the efficiency of the transportation system; to refine ways in which land is parcelled and exchanged; to improve the efficiency of the built environment so as to minimise waste in the production process. At the same time the construction and development industries are constantly seeking new markets for their products, calling for increased investment in suburbanisation, the reconstruction of existing urban areas and increases in building productivity (e.g. through the introduction of industrialised housing construction methods in the sixties).

The second process is the struggle between labour and capital over the distribution of resources. Not only does this relate to the distribution of surplus in the form of wages or profits (and therefore the amount that individuals have to spend on the consumption of housing, personal expenditure on retailing, leisure and mobility), but it is also manifest in the struggle over collective consumption (for example money to spend on environmental improvements and social infrastructure rather than investment to meet the needs of industry and commerce). Sometimes there is a coincidence between the needs of capital and demands for collective consumption and this type of expenditure, such as new shopping centres, car parks, commercial leisure facilities, tends to have a constant presence in urban renewal activity. In contrast collective consumption of non-profit making social infrastructure (such as social housing, health and welfare facilities) tends to fluctuate according to the changing relative strengths of capital and labour in the bargaining process.

In this chapter three types of urban renewal activity have been identified. The first is the continuing market led activity of adapting urban areas to the changing needs and demands of capital. The second takes the form of state social expenditure to ensure social harmony and well-being. For example the state's involvement in the renewal of housing and related social facilities, which has changed in nature from time to time but which, more importantly, has expanded in times of labour shortage (the 1950s) and contracted in times of labour surplus and weakness (the 1930s and 1980s). The third takes the form of state social physical capital investment and state regulation to facilitate profitable private sector property development and redevelopment. For example, the state's involvement in the wider process of urban regeneration, stimulating urban markets and preparing the basic framework within which private capital investment can take place, through such mechanisms as grants for derelict land reclamation, provision of utilities and public transport networks.

Certain events and trends can be seen as crucial to the understanding of the evolution of urban renewal. The first is the key role played by transport improvements in changing regional economic structures, changing land use patterns, reducing urban densities and facilitating rising housing standards. A second has been the consequences of manufacturing decline and service sector growth in changing the nature of demand for land and buildings within urban areas. A third, which in part derives from the first two, is the increasing difference between central and peripheral regions in terms of the economic conditions under which urban renewal is taking place, with rapidly rising demand for urban space in London, the South East and adjoining regions but generally slower growth, static or even falling demand for land and buildings elsewhere.

3 Economic Theory for Urban Renewal

To discuss urban renewal processes from an economic viewpoint requires contributions from a number of different strands of economic theory. The general demand for buildings derives from the needs of producers and consumers for urban space and their ability to pay for such space. Urban renewal comes about through action within the construction and development industries and will only occur under conditions which those industries find satisfactory or are prepared to tolerate. The urban renewal decision results from a view being taken about the economic life of existing buildings and the potential earnings or utility to be gained from a refurbished or replacement building. The value of existing and potential buildings depends upon the nature and workings of the urban land market at given points in time. Thus it is pertinent to examine in turn: the demand for construction; the supply of construction; the urban land market, the economic life of a building and the renewal decision. In this section the reasons and possibilities for state intervention in the renewal process and the economics of regeneration are also examined.

3.1 The Demand for Construction

The demand for construction has to be examined in two parts: firstly the demand for construction as an investment good, and secondly the demand for construction as a consumption good. Most construction other than housing is usually regarded as capital investment (factories, offices, shops, most transportation infrastructure and basic utilities) in that its only purpose is to contribute to the production process. Thus factories are used by industrialists in the manufacture of goods, offices provide space in which services may be performed, shops provide space where goods may be sold. Most transportation infrastructure is provided to reduce the costs of transporting raw materials and finished products and utility networks are also prerequisite to production.

The demand for investment goods behaves in significantly different ways from that for consumption goods because it is a derived demand. That is to say it is a function of the level of demand for the final product or service that is produced in the factory, office or shop. There is no inate demand for these premises other than for the space they can provide for production or realisation. Because of this relationship the demand for investment goods is influenced by what is known as the 'accelerator principle': as demand for any consumption good increases or decreases, then the demand for investment goods (particularly fixed capital) used in its production will increase or decrease at a greater rate.

The principle can be illustrated in the following example. Suppose there is a firm producing 10,000 items a year on five machines. Each machine is working to capacity and one machine is replaced each year. If there were to be an increase in demand for these items in one year of say, 10 per cent it would be necessary for the firm to purchase one whole additional machine to produce the additional output even though much of the capacity of the machine would remain unused. If demand for these items increased by another 10 per cent in the following year it would not be matched by another increase in the demand for machines. In fact there would be a decrease back to the previous replacement-only level until spare capacity was taken up. If demand for these items fell again then it is possible that even the replacement demand for machines would suffer if the firm anticipated that it could produce future levels of output on fewer machines. Table 3.1 illustrates this chain of events.

Table 3.1 The accelerator principle

Year	Output of pencils	% change in output	Investment in replacement machines	Investment in additional machines	Total investment in machines	% change in total investment in machines
1	10,000	–	1	0	1	–
2	10,000	0	1	0	1	0
3	10,000	0	1	0	1	0
4	11,000	+10	1	1	2	+100
5	12,100	+10	1	0	1	–50
6	12,100	0	1	0	1	0
7	12,100	0	1	0	1	0
8	10,000	–17.3	0	0	0	–100

Thus it can be seen how the demand for investment goods is much more volatile than is the demand for consumption goods. This relationship offers an important explanation for the nature of demand facing much of the construction industry. In practice these patterns are blurred as:

(i) there are tremendous variations in the rate of replacement as the existing built environment is composed of structures of varying age, quality, suitability and adaptability;
(ii) within certain general overall trends the demand for different consumption goods fluctuates at different times and by different amounts (for example whether the goods are for home consumption or for export);
(iii) speculative developers tend to try and anticipate rather than react to trends so that there are always a proportion who get it wrong, building where there is no demand and not building where there is.

However, generally at times of rising consumption one might expect to see initial sharp rises in demand for non-housing construction whereas in times of static or falling consumption, greater reductions in non-housing construction demand might be anticipated.

State intervention complicates this picture in a number of ways. The state plays a major role as the principal client for much fixed capital investment (roads, utility networks, land reclamation, advance factories, etc.) and it does not always time or locate such investment in accordance with market criteria. For example the state may use a change in its own fixed capital investment as a tool of macro-economic policy. It is attractive for governments to use construction as a national economic regulator for the following reasons:

(i) because it is much easier to make short run adjustments to capital spending than to reverse or expand ongoing revenue programmes and as construction activity represents a high proportion of public capital expenditure it is inevitably substantially affected by such changes;
(ii) the short run adverse political impact of cuts in construction activity is minimal since any one year's activity represents only a marginal increase in the total building stock of the country.

(Ball, 1988, p.5)

While urban renewal forms only a part of total construction output there is evidence to suggest that this proportion is increasing, and while construction output was cut back in the early 1980s as part of more general public expenditure adjustments the impact on urban renewal was somewhat cushioned. More recently, with an upturn in economic activity, there has been a sharp increase in some forms of urban renewal. Figure 3.1 shows the changing composition of construction output.

The indications of an increase in urban renewal are as follows. There is strong government encouragement for the little public sector housebuilding

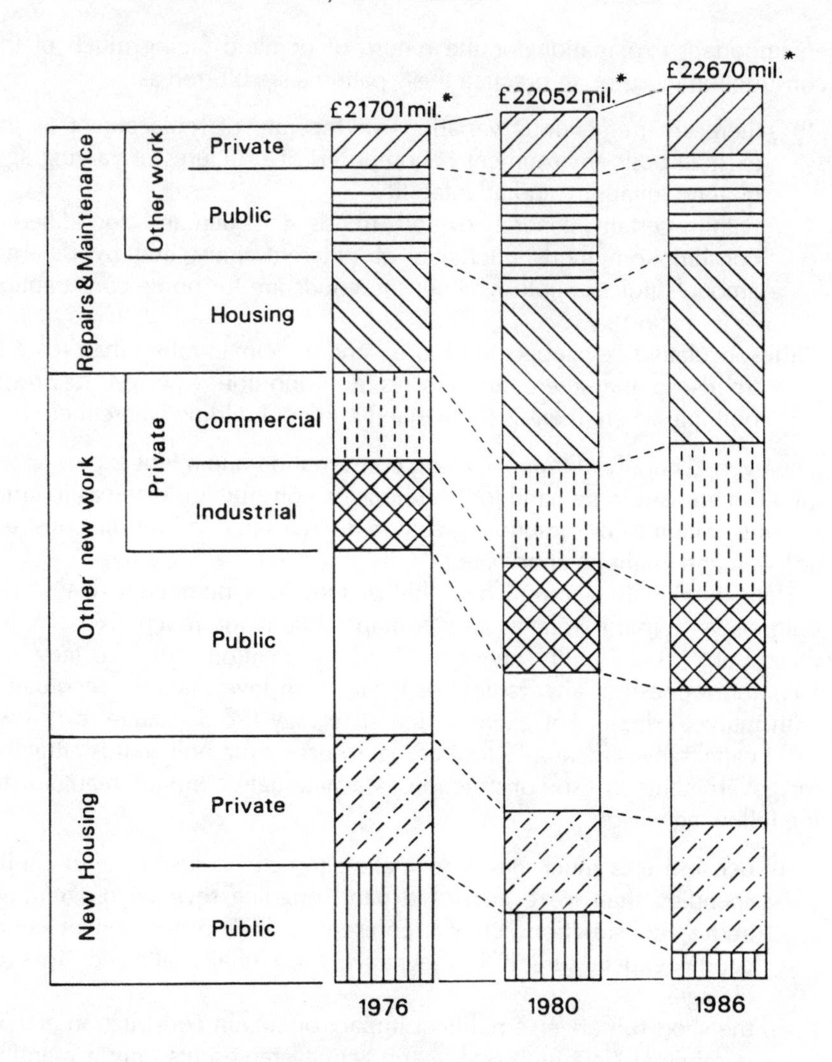

Figure 3.1 Changes in the composition of construction output 1976–1986

that it continues to permit to take place on redevelopment sites within urban areas and it seems possible that as much as three-quarters of council and housing association housebuilding is now occurring on these sites. While private housebuilding is less susceptible to government protestations there has nevertheless been a shift away from peripheral sites towards redevelopment and building conversion work in recent years. With other new work in the public sector it is difficult to identify any particular spatial location except that there is a general tendency for such activity to be associated with other forms of construction (either in providing pre-development infrastructure: roads, sewers, etc., or post-development infrastructure: schools, welfare facilities) and so to that extent it will tend to follow other trends towards urban renewal.

With regard to private industrial investment a number of trends are apparent. The first and most obvious is that there has been a marked decline and the second is a shift away from peripheral regions. Figure 3.2 shows some aspect of this regional movement.

A third point is that there has been a growing tendency for industry to decentralise and seek suburban or rural sites and to vacate and abandon inner city locations. This normally occurred through the birth of new firms or the opening of subsidiaries and branches on the periphery and the death or contracton of older firms in the inner city rather that through any singificant relocations of existing firms. As Fothergill and Gudgin have indicated, the lack of expansion space within existing urban areas provides a major explanation for this trend.

> The lack of room for expansion in cities will be felt most acutely by growing firms, who will be forced to relocate or divert some of their additional output into branches in small towns, or even forgo expansion altogether. In contrast, the lack of room for expansion clearly poses no problem to the firm which contracts or closes for reasons unconnected with location. We would therefore expect the location of additions to employment, in new factories and factory extensions, to differentiate cities and small towns to a greater extent that the location of job losses in contractions and closures.
>
> (Fothergill and Gudgin, in Goddard and Champion (Eds), 1983, p. 39)

Thus it may be assumed that much of this declining amount of building for industry is occurring in small towns and suburban locations and that relatively little takes the form of urban renewal. However, the trend is ameliorated by the array of government incentives now operating to some effect, encouraging private bespoke and speculative factory construction within urban areas: often taking the form of small advance factory units or managed workspaces within the boundaries of urban development corporations or inner city programme areas.

In contrast, the vast majority of commercial development (offices and shops) constitutes some form of urban renewal. Again there is a clear regional shift over the past fifteen years 1977 and 1987 and to some extent there has been a growth of out-of-town shopping and campus office

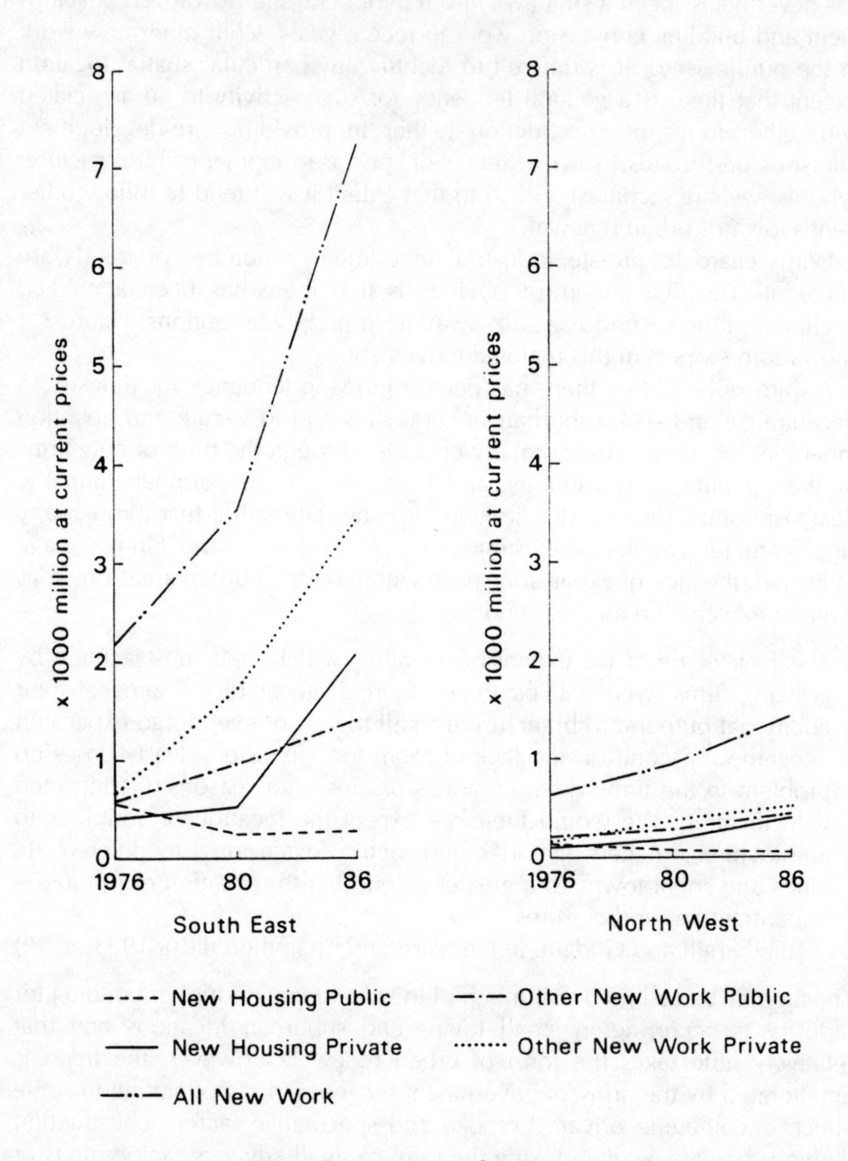

Figure 3.2 New orders obtained by contractors 1976, 1980, 1986. The North West and South East regions compared

developments. In spite of this, office building is still very concentrated in central area locations and therefore inevitably involved some demolition and replacement or conversion of existing property. While there has also been an upturn in retail investment lately a greater proportion of this has occurred in off-centre locations as some retailers have sought new types of outlets in the form of large single-storey retail warehouses with contiguous car parking and good road access. In many instances local authorities have been successful in encouraging such delelopers to take redundant former industrial sites within the existing urban area and these schemes have played a useful part in bringing such land back into beneficial urban use.

On the repair and maintenance side, which is entirely associated with urban renewal, we can see very clear patterns of growth through the whole period, except for the 1981 recession. Both the public and private housing stocks have been the subject of major refurbishment programmes with the benefit of substantial government subsidies. In the other sectors the increase in repair and maintance is mainly a reaction to two phenomena: one is simply the growth in the amount of building stock that exists and therefore requires maintenance, and the other is a gradual increase in the proportion of the non-housing built stock that has come into some form of corporate ownership and consequently more likely to be in receipt of regular planned maintenance than would have been the case under single ownership.

Thus it might be concluded that the amount of urban renewal and the proportion of construction that may be defined as urban renewal has increased in recent years. That is: a higher proportion of housing built on cleared sites; an increase in building for commerce which has mainly occurred in renewal situations; and increases in building maintenance, repair, renovation and conversion.

Furthermore, in recent years there have been fundamental changes in the context within which property development and urban renewal occur.

> The monetarist policy and ideology of governments since the late 1970s have wrought a major qualitative change as well as a simple shrinking in (non-military) state activity. The quantitive changes are familiar and well documented ... The qualitative change has been the enlarging of the sphere in which goods and services are traded as commodities, in which users become individualised in their 'market' relations with producers and providers.
>
> (Edwards M in Ball *et al.*, 1985, p.204)

Edwards goes on to cite examples of the commodification of housing through the sale of council housing and the reorientation of rent assessment towards market criteria; the concentration of Derelict Land Grants on so-called 'hard' (i.e. money earning) end uses; and even transportation investments, such as the Channel Tunnel, which have to be justified in terms of market returns. Insurance companies and pension funds have gradually come to dominate property investment and, as their funds continue to be swollen by government policies encouraging private insurance and personal pension schemes, their role is unlikely to diminish, although in future more funds may be diverted abroad. As a nation of owner occupiers and private pension and insurance policy holders we are increasingly dependent for our future well-being on the long run profitability of property capital. Since so much of the money flowing into these institutions is invested in property it becomes in everybody's long term interests to ensure that these investments remain valuable.

> Perhaps a third of the households in the country have debts secured against titles to their houses. These debts are to building societies and banks for whose investors they constitute valuable assets. Likewise in the non-residential sphere large sums stand in the balance sheets of investment institutions as the exchange value of the property assets they own. Even expert prediction of oversupply seems unable to stem the flow of new investment funds into office construction in London and speculative development in many parts of the country even in the crisis.
>
> (Edwards M in Ball *et al.*, 1985, p.211)

It has been suggested that this phenomenon has two consequences: the first is that the continuing confidence and steady growth of property values is now vital to the stability of the whole financial system; and the second is that these buildings are more and more required to be tradeable commodities. This means that these buildings must become more standardised in all their features so that bespoke buildings, unconventional designs and suspect locations are unlikely to be looked on favourably by these institutions.

The implications of this for urban renewal seem clear. The property investments of these financial institutions are already concentrated into fairly standard types of property: mainly city centre offices and retailing in London and the South East and a few 'safe' regional centres. There is every reason to suppose that this trend will continue, leaving less favoured locations and unconventional types of property to find alternative funding sources. In addition, it is possible that market activity and government support will be under pressure to ensure the maintenance of property values in these favoured locations, through the provision and maintenance of physical and social infrastructure and amenities, and given limited resources this is likely to be at the expense of support for infrastructure and amenities in less favoured locations. Thus we may well continue to see more government investment in the renewal of the built environment: (i) in parts of London and the South East rather than in peripheral regions; and (ii) in city centres rather than inner cities.

3.2 The Demand for Housing

The level of aggregate demand for consumption goods and services including housing depends upon the relative power of labour *vis-à-vis* capital in the struggle over the distribution of surplus at any one place and time. Thus during the 1930s a dual situation evolved. Skilled and professional workers, especially in the South East and West Midlands, were able to sustain a rising level of housing demand. At the same time, unskilled workers who were in surplus supply, especially in the North and North West, Scotland and Wales, were unable to sustain such rising demand but instead provided a cheap source of construction labour to meet these demands from elsewhere. During the era of relative labour shortage and buoyant output in the post-war boom years overall housing demand was again rising whereas from the mid-seventies onwards, during times of recession and labour surplus, housing demand fell substantially. It is important to understand that no distinction is made here between public and private housing since each can be seen as merely alternative mechanisms of response to labour demands for better housing.

Within this overall context it is then possible to discuss a number of factors that mould the character and sub-division of housing demand into certain tenures, locations and dwelling types. These factors include price, the price of substitutes, income and preferences. In the absence of state intervention the extent to which demand might be channelled towards urban renewal (i.e. dwellings on redeveloped sites, refurbished or converted dwellings) depends upon the relative price of urban renewal compared with new accommodation on greenfield sites.

The price of housing is a concept that is open to many interpretations. At its simplest, price may be regarded as the total selling price of an owner occupied dwelling, let us say, £40,000. However, if it were desired to compare the price of an owner occupied dwelling with that of a rented dwelling one approach would be to compare monthly outgoings: say, mortgage repayments of £400 compared with rent of £380. However, this leads to further complications since the monthly outgoings of the purchaser depend upon the price of the dwelling when it was purchased (which may have been long ago), the size of deposit put down by the purchaser and hence the size of mortgage, the type of mortgage, the period of loan and the rate of interest payable. Thus for identical dwellings currently commanding the same value, the price paid (in monthly outgoings) could vary considerably.

These differences in circumstance are important because they indicate not only the amount of equity or financial security enjoyed by each owner but also their relative ability to undertake building maintenance, repairs and refurbishment. In a period when the government has been striving to increase the proportion of owner occupied dwellings and where many owners have been encouraged by lenders to borrow to their credit limit, it is only the inflation of house prices that creates the availablility of funds for repairs, etc. Having now (in 1990) reached a plateau in house prices after a major growth period there is the real possibility of a starvation of maintenance funds for much recently acquired property.

If prices are to be compared over time then inflation has to be accounted for, usually by indexing house prices against average price changes (e.g. the retail price index) or against average income levels.

The perceived price of a dwelling may also reflect such items as the cost of transport to work, which is a necessary element of housing costs for all employed people. Price may also be subjectively assessed against different levels of local amenity and the character of the neighbourhood.

Prices may be distorted by subsidies. For many years the 'price' of public sector housing was held below market levels by central government supply side subsidy. This had the effect of stimulating housing demand and skewing it towards the public sector. Some local authorities supplemented this with their own additional supply side subsidy derived from the local rates. This again further distorted demand and, where a conurbation was split between more than one local authority (as in London) it added a spatial effect, increasing demand in some boroughs more than others. The effect of rent control has already been discussed: reducing housing supply and increasing demand through, in effect, requiring landlords to subsidise tenants. Both these distortions have had major long run implications for housing renewal as both the private rented sector and many local authorities (partly through their own choosing) have either not thought it economic or have been starved of funds to undertake an adequate level of housing maintenance and repair. As a result much of this housing has fallen into a state of decay where major renovation or demolition become the only solutions. Without these distortions in the market the level of housing maintenance might have been higher, there might have been less general decay in the stock and the nation might not be facing the acute housing crisis of recent years. The evidence for this view comes from West Germany where, in a stock that is predominantly privately rented, it has normally been accepted that the primary function and requirement of rent is that it must cover all the landlord's costs. The consequence of this system has been a high level of investment in both new construction and maintenance of the sector and a relative absence of the physical decay that characterises so much of our inner city housing stock (Couch, 1985, p 82).

This is not to make any political judgement as to whether, to what extent, or how the price of housing should be subsidised but simply to raise the issue that the type of subsidy system used in this country over a very long period of time has probably imposed significant costs in terms of the resultant quality of the housing stock. Both these types of subsidy are being reduced or phased out by the present government as it seeks to eliminate public sector supply side subsidies and to return private sector rents to market levels. These changes of policy, of course, impose other personal and social costs.

Another type of subsidy that has become increasingly important is the income support or demand side subsidy. The two basic forms are housing benefit paid to low income groups to enable them to pay housing rents and tax relief on mortgage interest payments paid to owner occupiers. It is worth examining the distorting effects of this tax relief system.

While the effect of tax relief is to increase housing demand, two points should be borne in mind. Housing supply is generally regarded as inelastic but secondhand supply is probably more elastic than new supply; and demand is not noticeably discriminatory between new and secondhand supply. Therefore in the new housing market the effect is to slightly increase the quantity supplied but to substantially increase price. This increase in quantity may take the form of more dwellings or bigger/better dwellings or both. The increase in price creates bigger development profits which are shared between builder/developers and landowners according to their relative bargaining strengths.

For secondhand housing the consequence is to increase the quantity supplied and price by more similar (but not necessarily equal) proportions. Here an increase in supply means a rise in the number of existing dwellings being offered for sale. Each time supply is turned over a profit is likely to be realised by the owner and the various professionals involved in the transaction. These profits made by owners can then be spent on other goods, saved or ploughed back into house purchase or renovation.

The level of demand for housing is also influenced by population. First and foremost demand depends upon the size and growth rate of the total population but this is substantially modified through a series of adjustments that take account of age structure, income levels and lifestyles to give household size and structure and so determine the number and types of dwellings that will be demanded by any given population. The elderly and childless are more likely to demand conversions or redevelopment of large properties into small easily managed flats while young families will tend to demand houses with gardens and safe local environments for children. However, it is important to remember that the influences of age structure and lifestyle are not independent of income. The late twentieth-century trend towards smaller households has as much to do with rising real incomes as with changes in the age profile of the population or unconstrained shifts in social behaviour. Neither is demand free from the influence of suppliers as they constantly seek new markets for their ever increasing capacity.

3.3 The Building Industry and the Supply of Urban Renewal

Urban renewal is produced through the construction industry: an industry that has certain basic features that differentiate it from most other industries and give a particular character. Today the industry itself is responsible for around 6 per cent of gross domestic product, and materials producers and other suppliers account for a further 6 per cent, making it by far the biggest single industrial sector in the economy. Recent figures indicate that around 1.5 million people work in the industry, including the self employed, and that there are around 175,000 building firms.

The number of firms has more than doubled since 1977, mainly as major firms have responded to uncertainty and the need for flexibility by shedding directly employed labour and replacing it with sub-contracted labour which can be more easily laid off at the end of a production phase. But a secondary reason for the increasing numbers of small building firms has been the growth in repair and maintenance work, including renovation and conversion projects. Much of this work takes place in small contracts using traditional building methods. This is a market in which economies of scale seem to be slight and the smaller firms appear well able to compete against the major companies. However, it is not the case that all renovation and conversion work is dominated by small firms:

> New opportunities have opened up for the major contractors in the large-scale urban renewal schemes usually financed through partnership arrangements of various sorts; they have also provided opportunities for diversification to the large speculative builders. In Liverpool Crudens have benefitted from large contracts such as the redevelopment of the Cathedral site and the refurbishment of large local authority housing estates. Barratts (Urban Renewal) have redeveloped Minster Court, Wapping Dock Warehouse and Stockbridge Village. Wimpey and Tarmac have large renovation contracts in various parts of the city ... In fact there are clearly two different markets within 'repairs and renewals'; there is small-scale work which has always been the bread and butter of the small firm sector but which has continued to expand during the recession as owners have chosen not to move to new premises; and there are the relatively new large-scale renewal programmes ... which seem to have become the province of the large firms.
>
> (Couch and Morton, 1988, p. 135)

There are further features of the construction and urban renewal market that influence supply. Even today after several years of intense privatisation by the government much of the market for urban renewal is still under the control or influence of central government. The government is virtually the sole consumer of civil engineering work, although there are moves to change this (e.g. the Docklands Light Railway), and as such it is a monopsony purchaser capable of pushing prices down. In many other instances, such as local authority and housing association new-build and refurbishment activity central government has a substantial influence on contract price and conditions.

Many urban renewal markets are very imperfect and subject to great distortion of price away from perfectly competitive levels. Sales in urban land markets are infrequent, differences between parcels substantial and secrecy about prices commonplace. Many central area partnership re-developments are negotiated between a public sector agency and a single developer. In most large refurbishment contracts only a handful of selected firms are invited to tender. Even in the market for new private housing most prospective purchasers are faced with only two or three choices of developer once they have selected a general location, price range and dwelling type. Only in the secondhand housing market and the market for small-scale maintenance, repairs and renovation do we move towards a more perfect market structure.

It is debatable whether there are any significant on-site economies of scale in urban renewal schemes. Most urban renewal schemes are one off bespoke projects, often on difficult and cramped sites. It is only at the level of the firm that scale economies are likely: for example through bulk ordering for a number of separate contracts, through the division and specialism of managerial labour, through access to cheaper sources of contract period funding.

Firms undertaking speculative projects will be structured differently from contracting firms. They make their profits through 'buying cheap and selling dear'. Critical to their success is their commercial acumen in site selection, land negotiation, market understanding, timing of development and marketing skill. Contractors on the other hand:

> do not instigate development but only build projects for clients. This creates two crucial distinctions with speculative building. First, building production becomes the means through which profits are made rather than a prerequisite for realising a development gain. Second, there is a division of control between the conception of the project and its productive implementation. The client is the developer, the contractor is the producer, and generally some professional designer will draw up the client's plans.
>
> (Ball, 1988, p. 48)

A further important characteristic of the production of urban renewal is that both building contractors and speculative firms are, in varying degrees, flexible and indifferent as to the precise nature and location the building services or development that they supply. That is to say that few contractors are limited by expertise or capital equipment to producing any particular type of building or sector of the market, although depending on size and inclination firms are to some extent restricted in the location and scale of project they can undertake.

Because of this any urban renewal project must be at least as profitable to the construction firm as the alternative outlets for the firm's productive capacity, or by and large it simply will not be built. Thus for the small local jobbing builder specialising in repair and refurbishment contracts operating in one city, an inner city improvement grant job must be at least as profitable as a suburban extension, and likewise housing sector work must yield at least as good a return as industrial or commercial work. For the major national firms with contracting and speculative divisions there will be a tendency to shift resources within the firm to whatever is the most profitable activity: from contracting to speculation or vice versa, from new build to refurbishment, from housebuilding to commercial and industrial development, or from depressed peripheral regions towards properous central regions. In conglomerate firms (which typifies many of the largest firms such as Comben Homes, part of the Hawker Siddeley group; or Bovis Homes, part of P & O Ltd) the building division (including whatever part of that is engaged in urban renewal) must maintain a level of profitability at least as good as those of other divisions.

3.4 The Economic Life of a Building and the Timing of Redevelopment

At the heart of understanding urban renewal is the idea of the economic life of a building and the timing and nature of the decision to refurbish, convert, replace or abandon it. For this is how urban renewal happens. Individual consumers, firms or the state make these decisions and initiate some kind of urban renewal process: some modification to the existing urban structure.

The economic life of a building can be considered as the period of time over which the capital value of the building exceeds the capital value of the cleared site. According to Goodall (1972) a building will not be demolished so long as:

$$Be > Sn$$

where *Be* is the present capital value of the real property and *Sn* is the value of the cleared site. *Sn* is determined as follows:

$$Sn = Yn - Cn - On - De$$

where *Yn* is the present capital value of the expected earnings from the replacement building, *Cn* is the cost of constructing the replacement building, *On* is the present value of the operating costs of the new building, and *De* is the cost of demolishing the old building and preparing the site. Also

$Be = Ye - Oe$

where *Ye* is the present capital value of earnings from the existing building and *Oe* is the present value of the operating costs of the existing building.

Making redevelopment decisions in the market involves the handling of a large number of variables, some of which are difficult to estimate and/or critical to the outcome of the calculation. Valuers may have to make assumptions about the permitted replacement use and the intensity of that use. They are concerned with the anticipated rent that can be achieved for such a building and with estimating maintenance costs, repair intervals, management and energy costs many years hence. They have to estimate demolition and site preparation costs (simple in a housing area, problematic in an industrial zone). They have to decide on an appropriate discount rate with which to bring all future costs and earnings back to a current value and they have to make assumptions about the rate of inflation in each of these costs and earnings (none of which may necessarily inflate at the same rate). This decision-making process may be simple and straightforward for a local speculative builder replacing a Victorian villa with a new block of flats, or, for a major insurance company with millions of pounds' worth of property assets it may be a highly sophisticated process using computer models to test a large number of decisions against a wide range of assumed possible conditions for every variable in order to select the preferred course of action. The economic life of a building is shown graphically in Figure 3.3.

The lifespan of a building is determined by the economic imperatives of the market or by social intervention, and outside of this context there is no definable 'physical' life of a building. That is to say it is not possible to make such a statement as 'typical British houses have a physical lifespan of eighty years', for it simply is not true. There is no predeterminable physical lifespan for a building outside of the economic and social context in which it exists. Thus in Britain it is possible to see dwellings up to 400 years old still providing perfectly adequate accommodation, or churches up to 1,000 years old still in their original function. Equally, there are new buildings in the London Docklands that are being demolished to make way for more intensive developments even before they have been occupied, and there are many thousands of local authority flats built in the 1960s that have already reached the end of their economic life and are being demolished. Thus the timespan in Figure 3.3 may be months or hundreds of years.

Leaving aside social factors for the moment, the market will determine the economic life of a building as ending when lines *SS* and *CC* intersect and, as such will depend upon the relative slope of these two lines over time. Line *CC* represents the price that the building commands in the

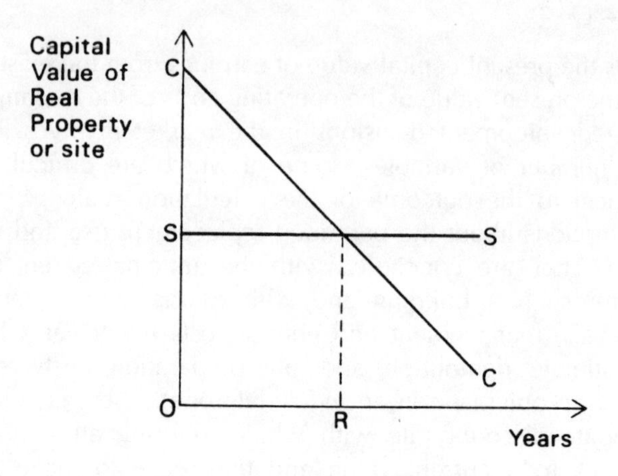

Where SS represents the value of the cleared site over time and CC shows the capital value of the developed site with a building

R = redevelopment

Source: after Goodall B, 1972, page 209

Figure 3.3 The economic life of a building

market minus the operating costs. The prices will be determined by the market and will vary in the same way as would the price for any other good. However, a problem arises in that what is normally being demanded is 'urban space'(i.e. developed or transformed land, the site with the building) and it becomes very difficult to separate what is happening to the price of the building from what is happening to the price of the land. A change in the level of demand for 'urban space', such as might be created by an increase in population or economic activity, will tend to push up the price of sites and buildings in the area. Thus both lines *CC* and *SS* will shift upwards at the same time. Whether they will shift to the same degree depends upon individual circumstances. In general a simple increase in the demand for housing across the board will tend to enhance the value of existing dwellings and prolong their economic life. However, if there were to be a change in the nature of demand away from family houses towards accommodation for small, childless households then the price that large houses could command might fall in relation to site values and the houses might be replaced by smaller flats. Similarly at the edge of the commercial centre of an expanding town an initial rise in demand for office space might result in a simple change of use of existing houses into offices, yet if demand continued to increase it might become profitable to increase the intensity of use by demolishing the existing 'houses' and replacing them with a purpose built office which would use the same site much more intensively.

Harvey (1981, p. 93) approaches the matter in a slightly different way. When a building is new it is generally the case that the income will be high in relation to the operating costs. As time passes there are tendencies for the real level of income to fall (as the building becomes obsolescent in relation to modern needs and patterns of building use, or if other more appropriate buildings are being supplied elsewhere in the market), and for operating costs to rise in relation to income (as the building ages maintenance costs will tend to rise, refurbishment may be necessary, voids may become longer and more frequent), until eventually operating costs exceed the income. At this point the current use may cease. The building may then be turned over to another more profitable use, perhaps after modification or conversion, or it may be demolished and replaced if the returns on the redeveloped building exceed the costs of demolition and reconstruction. If none of these possibilities is financially viable then the building may be abandoned and left vacant (i.e. if it has a negative value in its present use, a negative value in any alternative use, and a negative value after redevelopment). In these circumstances nobody will have any financial reason to invest money in the premises and dereliction will tend to follow.

It will be clear from empirical studies of dereliction in Britain just how many sites do in fact have negative values. In 1982 the Department of the Environment identified some 45,683 hectares of derelict land in England (DOE, 1984). Although the official definition of derelict land: 'land so damaged by industrial or other development that it is incapable of beneficial use without treatment' makes no reference to financial calculations the essence of the problem is that the costs of site clearance and restoration plus the cost of building far exceed the likely return from such investment. Thus the function of the Derelict Land Grant and, where it is used for similar purposes, the Urban Programme, is to subsidise restoration costs so that redevelopment becomes a feasible proposition for the market to consider.

The possibility of a site being redeveloped depends upon its yielding not only a positive return (a profit) but also at least as much profit as the alternative possibilities available to potential investors. This applies to all market sectors but is perhaps most starkly illustrated in relation to the office market, in which a relatively small number of London based financial institutions dominate decision making. The hypothetical residual valuation in Table 3.2 illustrates the contrasting attractions of investing in London or Liverpool.

Table 3.2 Hypothetical financial appraisal

	Liverpool	*London SW1*
Anticipated rent level	£10 per sq.ft	£70 per sq.ft
Anticipated net rental income for 20,000 sq.ft lettable floor area	£200,000p.a.	£1,400,000p.a.
Years purchase	12.5YP	20.0YP
Capital value after development	£2,500,000	£28,000,000
Development costs of 20,000 sq ft	@ £80 per sq.ft	@ £100 per sq.ft
Total	£1,600,000	£2,000,000
Building finance for a two year contract @ 12% for ½ construction period	£192,000	£240,000
Professional fees, etc.	£320,000	£400,000
Total costs	£2,112,000	£2,640,000
Residual profit available for land purchase	£388,000	£25,360,000

The distribution of this residual profit between the landowner, in the form of land price, and the developer, in the form of developer's profit, will depend upon the relative power of each in the market at the time the deal was made. In circumstances such as Liverpool, where the demand for office floorspace and therefore the demand for development sites is low, then the developers are likely to be in a stronger position *vis-à-vis* landowners and so be able to extract a proportionately higher rate of profit by paying a relatively low proportion of the residual to landowners as the land price. In a more vigorous market, such as London, where the demand for office space is very high but the supply of suitable developable sites is limited, then the landowner will normally be in a stronger position and be able to extract a higher proportion of the profits by forcing the developer to pay a relatively higher price for the land.

Let us assume, for the sake of argument, that the Liverpool landowner extracts 40 per cent of the residual as the land price while the London landowner extracts 60 per cent. The results would be as follows, giving the developer a possible 10.2 per cent return on their investment in Liverpool but a possible 56.8 per cent return on the London development.

	Liverpool	*London SW1*
Land price paid	£155,200	£15,216,000
Developer's profit	£232,800	£10,144,000

The relative attractions of different forms and locations of property investment are sensitive to changes in any of the variables in this calculation: changes in local rent levels, building costs, interest rates, fee levels and land prices can be critical in determining the likelihood of one particular development going ahead rather than another. Indeed, in peripheral cities it is not uncommon for the residual value of the site to be negative and for no redevelopment to take place without subsidy. The market will normally tend to redevelop the most profitable sites first and only when the prime development opportunities are taken up will potential investors turn to secondary sites (whether in secondary zones of London or the primary zones of other cities). The situation may be compared with the Ricarian theory of agricultural land rent, where the most fertile land is used first and only if more land is needed will land in the next lower grade of fertility be brought into use. Likewise in property investment, only if all the prime, highest yielding opportunities are taken and if the financial institutions still have money that they need to invest, will they consider secondary opportunities.

This is not only true for the commercial (office and retail) market but also for speculative industrial developments and private housing projects, although the latter are usually funded from different sources. This distortion of the national property investment market is only ameliorated at the margins as (i) some financial institutions indulge in some spatial risk spreading; (ii) there are token regional investments in acknowledgement of social obligations; (iii) the state sometimes intervenes to reduce development costs and risk in peripheral locations through various forms of subsidy or guarantees.

This clearly leaves investment opportunities and development requirements in peripheral towns and cities, such as Liverpool, which are not, and may never be, funded through the major financial institutions. In some cases these opportunities may be taken up by local entrepreneurs who are not big enough to compete in the prime London markets. These entrepreneurs will use different sources of funds and may accept lower and more risky returns than the major institutions. But the network of local entrepreneurs is not well developed, except in the private housing market, and they will probably not take up all the opportunities that exist. In addition some redevelopments will occur as building users commission bespoke projects for their own needs, financed out of their own profits or through loans set against the collateral provided by their business or property assets.

3.5 The Economics of Urban Vacancy and Dereliction

According to a recent literature review on vacant urban land:

> The biggest single research gap is that we do not know whether the vacant land problem is getting worse, or better, or staying the same.
>
> (Cameron *et al.*, 1988, p. 119)

Some progress has been made in recent years in recognising the need for data about the incidence of vacancy. The main sources of such data are the Department of the Environment's own Derelict Land Surveys and the Registers of Vacant Public Sector Owned Land. Cameron *et al.* consider the various causes of vacancy suggested by previous researchers and produce a long list of possible causes, but they conclude that much of the existing analysis lacks an adequate theoretical base:

> we have found in this report that a whole host of different classifications of the causes of vacant land problems have been used. So we need to strive towards a more consistent approach to investigating the causes of land vacancy. The unemployment analogue has perhaps as much potential here as any other. We might ask whether vacancy is qualitative, frictional, demand deficient, cyclical or structural.
>
> (Cameron *et al.*, 1988, p. 120)

While it must be acknowledged that there are significant differences in the nature of land and labour markets, as factors of production the way in which they are demanded and used in production is sufficiently similar to justify the analogy. Thus in Figure 3.4 the causes of initial and continuing vacancy are explained using the unemployment analogy.

Increases in demand for goods and services bring land into urban use. Once in this category there are various events that can lead to a decline in intensity or total cessation of use. (Intensity can increase but that is not our concern here.) Firstly it is possible that a change in the means of production by the firm occupying the land could lead to vacancy; this might occur if, for example, a new technological process required less space in which to manufacture the same level of output. In this situation and assuming no change in other parameters such as demand for the product, the firm would have no further use for the vacated land and would then normally attempt its disposal. If a new occupier were quickly found the

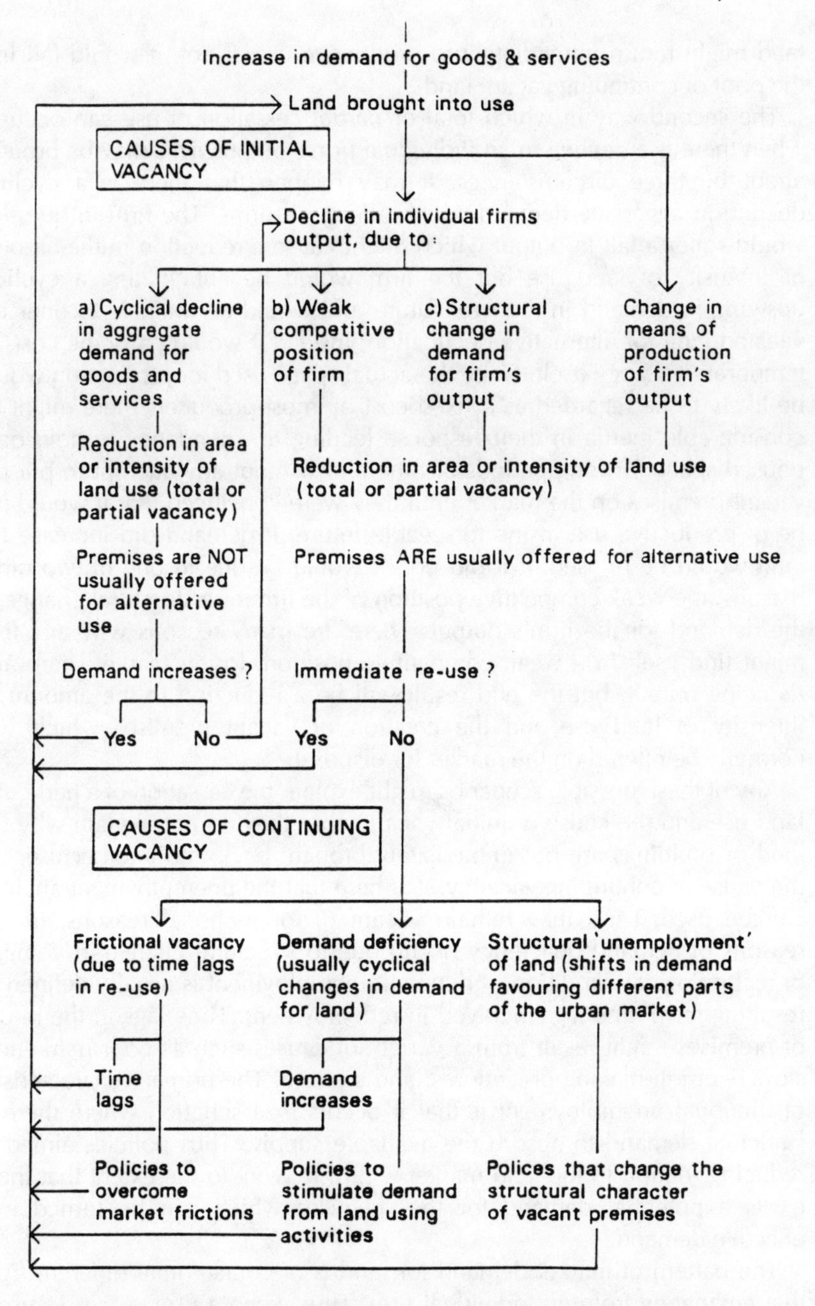

Source: after Couch C et al Urban Vacancy in Liverpool in Grover R (Ed) 1989 Land & Property Development: New Directions E & FN Spon, London.

Figure 3.4 The causes of initial and continuing vacancy

land might return immediately to productive use, if not, it would fall into thc pool of continuing vacant land.

The second way in which total or partial cessation of use can occur is when there is a decline in an individual firm's output. This may be brought about by three circumstances. It may happen that there is a cyclical decline in aggregate demand which affects all firms. The firm in question would suffer a fall in output which might cause a reduction in the amount or intensity of land use but the firm would be anticipating a cyclical upswing in demand in the near future and would be unlikely to offer the vacated land for alternative occupation; instead it would carry the costs of temporary vacancy itself. Since the actual space used for production would be likely to be regarded as a fixed cost by most producers there might be considerable inertia in their response leading to a tendency to hold onto unused space for long periods of time and without any attempt to put the vacant premises on the market until they were convinced that it would not be of productive use in the forseeable future. If demand did increase the land would be re-used; if it did not, it would be due to one of two other reasons: the weak competitive position of the firm or a structural change in the demand for the firm's output. There are many reasons why any firm might find itself in a weak competitive position, losing market share and reducing output, but the end result will be a reduction in the amount or intensity of land use and the creation of vacant premises which will normally be offered on the market for disposal.

Any of these possible scenarios might explain the cessation of a particular land use and the cause of initial vacancy but they do not explain why the land or buildings are not immediately brought back into another use, i.e. the cause of continuing vacancy. It is here that the unemployment analogy can be used. Land may remain vacant (i) for frictional reasons, (ii) for reasons of demand deficiency or (iii) due to structural change in demand or technological conditions. Frictional unemployment is usually defined as resulting from time lags involved in redeployment. Time lags in the re-use of premises might result from a variety of causes such as poor marketing, slow decision-making procedures, and so forth. The primary characteristic of frictional unemployment is that it occurs in a situation where there is sufficient demand to absorb the available supply. Thus policies aimed at reducing friction in the land market will only work to the extent that there exists a potential demand for the premises which can be turned into effective demand.

The pattern of market demand for land is of course quite different from that emanating from an individual firm. Thus when a firm ceases to use a site because of structural changes in its means of production or because of its weak competitive position in the market this says nothing about the general demand for land. If the market demand for land is buoyant the site is likely to be speedily re-used whereas in periods of slack demand vacancy is more likely to continue. Even in circumstances where land is underused because of cyclical decline in the aggregate demand for goods and services affecting all firms, it does not follow that any vacant land that is released onto the market will necessarily face a sluggish market: the movement of building cycles does not correspond well with trade cycles and the land market even less so.

Nevertheless some land will remain vacant, not for frictional reasons nor because of changes in the nature of demand (see below) but simply because it faces a low level of demand brought about by cyclical movements in the land market.

Structural unemployment occurs when there is some basic long term change in the nature of demand or technological conditions. With regard to the 'unemployment' of land, Cameron *et al.* suggest that:

> The structural under-utilisation of land may, in this context, be caused by two different phenomena, the first being the long run substitution of land by capital so that the land to output ratio falls; the second being a shift in demand favouring different areas of the urban land market. Under the first condition a given volume of urban output can be achieved with less land than previously so that some existing supply is not taken up. Under the second condition some areas of the urban market become preferred locations and other areas are seen as inappropriate. Here again the marginal effect is a persistent under-utilisation of land available for use in specific locations.
>
> (Cameron *et al.*,1988, p. 120)

The first of these phenomena is more useful as an explanation of cessation of use or initial vacancy. It is clearly not the case that the land to output ratio is falling in all market sectors (consider for example the differences between the increasing density of city centre office developments and the decreasing density of out-of-town hypermarkets and shopping centres) and therefore as an explanation of continuing vacancy it is only likely to be of use where sectors of the land market are discrete and not subject to competitive demand from other sectors: a rare situation.

The second phenomenon can be perceived in British cities where vacancy has been the result of:

(i) changes in locational preferences, e.g. the increasing attraction of proximity to motorway junctions and the declining attraction of proximity to railway access or, the greater importance attached to local environmental conditions rather than access considerations for some types of industry;

(ii) changes in the type of premises demanded, e.g. the rejection of very small sites, buildings of awkward shape or lacking in certain characteristics.

Thus some continuing vacancy can be explained by these structural changes in the land market.

3.6 Justifications for State Intervention in Urban Renewal

Given the potential for unsatisfactory outcomes in urban renewal if the market behaves in the ways discussed above it is not suprising that the state does intervene in urban renewal processes; but in what circumstances does it intervene? What criteria does it use to determine priorities? How does it justify its interventions?

Economic theory provides a number of different reasons for state intervention in urban renewal and these can be identified through both Marxian and neo-classical perspectives. O'Connor (1973), a Marxian economist, formulated a complex sub-division of state expenditure which is relevant here (although intervention in urban renewal includes both state expenditure and regulation).

He begins by suggesting that state expenditure can be divided into two basic categories: the legitimation function and the accumulation function. The legitimation function requires that the state incurs social expenses to achieve and maintain social order: such expenditure includes the provision of defence and police functions and social policies, often of a token nature, designed to avoid public unrest or disorder of one sort or another. The accumulation function of the state is more complicated but more directly related to assisting the good working of the capitalist economy. Here the state provides social capital, that is to say all the commonly owned infrastructure upon which the capitalist system of production depends. O'Connor sub-divides this into social consumption, which includes goods and services collectively consumed (e.g. health services) and social investment, which he further sub-divides into investment in human capital (e.g. education) and physical capital (e.g. transportation and utility networks).

Under this kind of analysis the logic of state intervention in urban renewal becomes apparent. Thus the accumulation function of the state (i.e. lowering the costs or increasing the efficiency of capital) might include the renewal of the inner city transportation networks, utilities, land reclamation and land assembly all as part of social physical capital investment. Expenditure on retraining schemes for the long term unemployed would be social human capital investment, although there may well be an element of social expenses here as well. Inner city environmental improvements might be a necessary prerequisite to new industrial investments (social physical capital investment) but might also be seen in part as necessary social expenses incurred in the maintenance of social harmony. However it should also be remembered that the market will only not undertake, or not seek to undertake, any of these functions while they remain unprofitable to any individual capitalist enterprise. Thus the boundary of what can be regarded by capitalists as the acceptable role of the state in urban renewal is constantly shifting.

Taking a neo-classical approach state intervention is categorised in different ways although much of the same ground is covered. Under this approach reasons are provided to explain how the state could justify intervention in various circumstances but the shortcoming of the neo-classical explanation is that it does not explain why the state does intervene in any particular situation.

The first reason is the divergence between the private and social costs of production or consumption of some good or service. The social costs of an activity (be it production or consumption) are those costs borne by society as a whole as distinct from private costs which are only those costs which are reflected in the market price of goods and services. Thus, for example, the private costs of detergent production in Warrington are the costs to the firm of the labour, materials, energy, etc. used in production, whereas the social costs would include the costs of pollution, traffic congestion and deterioration of public roads, etc. caused by this production. Thus it is argued that an important reason for state intervention is to deal with this gap between private and social costs. It is suggested that the producer should be taxed or subsidised or regulated in order to encourage or to compel account to be taken of social costs. The logic of this intervention is that since the social costs are borne by society and are therefore a burden to the whole economy then intervention to reduce them will release resources that could be used for more useful purposes. The limitation of this view is that it assumes that the state wants to deal with pollution. The Marxian view would be that the state will only intervene if it is in the general financial interests of capital or if it is forced to do so through social pressures.

Secondly neo-classical economists point to market imperfections as a reason for state intervention. There are a variety of reasons why markets such as the land market or the housing market may contain imperfections: inertia amongst producers or consumers in responding to changes in demand, supply or price; lack of knowledge about aspects of the workings of the market; the distorting effects of monopoly or monopsony power and so forth. The state may intervene to ease these difficulties and distortions in order to improve the functioning of the market.

Some goods and services are necessary parts of the infrastructure of capitalist society but are difficult or impossible for the market to provide on an individual consumption basis. Examples include much of the physical infrastructure of cities: roads; utilities; social control and protection services; parks and recreation areas. Since each individual desires to consume some small proportion of these items but since it is difficult or impossible for individual consumption to be transacted then it is a logical response for some central agency, usually the state, to purchase or consume such items on behalf of the community.

Merit goods are said to be provided by the state because political value judgements are made that certain groups of people should have these goods or services regardless of their ability to pay or their own personal desires in the matter: classic examples are education and health care for children. Given that a market economy inevitably produces great divergences in income levels and the ability to pay for goods and services it is argued that one further reason for state intervention is to redress some of the imbalance created and to 'help those least able to compete in the market' to do so, or at least to obtain the basic necessities of modern life. Certainly transfer payments through the state and state subsidies (e.g. housing benefit) are effective ways in which the state does achieve such aims but there is again nothing in neo-classical economics to explain why the state should make such interventions. From the Marxian point of view it would be argued that the logic for such expenditure can be found under two headings: social expenses required to maintain social harmony and social investment in human capital (reproduction of the labour force).

Given that the state decides to intervene in urban renewal processes there is a very limited number of approaches it may adopt:

(i) rules or regulations: which may be positive or negative, permissive or mandatory;
(ii) expenditure; capital spending (the purchase of fixed assets); revenue spending (including subsidies and tax foregone);
(iii) income; taxation and the pricing of state services;
(iv) advocacy.

Thus the state may make rules or regulations (i.e. Acts of Parliament, statutory instruments and bylaws) regarding any aspect of urban renewal. These may be positive, indicating that something must be done, e.g. conditions on a planning permission requiring that parking places are provided or that landscaping must be undertaken; or they may be negative, e.g. forbidding the demolition of a listed building or the location of an intrusive factory in a residential zone. They may be permissive, as was much nineteenth-century public health legislation which allowed local authorities to take certain kind of action (e.g. slum clearance) if they so wished; or they may be mandatory, i.e. imposing a duty upon some individual or agency to ensure that in given circumstances certain action will follow, e.g. under the 1980 Housing Act local housing authorities must permit sitting tenants to purchase their council dwellings.

State expenditure may also be used to achieve its aims either in addition to or instead of the making of regulations. Thus the listing of buildings may be backed up by making grants available for their restoration. The state can spend money directly itself, or through its own subsidiaries – the local authorities and statutory undertakers – or it can subsidise others to carry out such work. Thus the state can build houses itself (as it used to in new towns), or finance building by local authorities or housing associations or subsidise private sector building (e.g. through the use of Derelict Land Grants). Tax foregone, as in mortgage tax relief or rate relief in enterprise zones, should also be regarded as state expenditure since it is a cost to the state.

State income can also be used to regulate local economies and achieve desired aims. Pollution can be reduced through taxation (or by subsidy or regulation) and the pricing of all goods and services sold by the state can have significant effects on demand and markets generally. Thus increases in council housing rents will encourage purchases by sitting tenants and promotion of the owner occupied market; rent free periods on new advance factory estates will encourage demand and their early occupation.

Beyond these approaches the state can also seek to extend its power through advocacy: propaganda, publicity and persuasion. Examples are: advertising the benefits of relocating firms to peripheral locations; offering advice to small businesses; offering 'strategic guidance' in the preparation of Development Plans.

3.7 The Economics of Regeneration

To generate is, according to *The Concise Oxford Dictionary* 'to bring into existence, produce, evolve'. To regenerate is to 'generate again, bring or come into renewed existence'. Thus when we talk of urban regeneration we are talking of something more than urban renewal: reuse and reinvestment in the physical structure of existing urban areas. We are talking about an economic process that increases what has decreased: of increasing investment where it has declined, of increasing employment where it has declined, increasing consumer expenditure and increasing population; in essence we are talking about economic growth.

For an entire country (a bounded economy) it is perhaps possible to identify a fundamental aim of economic regeneration as sustained long run economic growth. That is to say: rising national income and growth in capital stock, employment and consumption. At its simplest national economic growth is achieved through increases in productivity (output per person) and/or through maintaining a surplus in the international balance of trade.

For an urban area economic regeneration is not necessarily achieved in the same way. In the first place urban growth does not automatically lead to urban regeneration: growth can take place at the periphery with little or no benefit to the declining parts of the city. Urban regeneration is much more dependent upon locational and distributional decisions than national economic regeneration. In the second place urban areas are not bounded economies so that the multiplier effects of investment and job creation are more susceptible to leakage out of the urban area than is the case with national economies.

Suppose an additional £1 million worth of goods is purchased from the construction sector of the national economy. It is possible that the construction sector might spend this money as follows: 50 per cent on the direct employment of construction labour, 40 per cent on raw materials, part finished goods, energy and services produced in this country, and 10 per cent on imported goods and services. In these circumstances 90 per cent of the original expenditure will have beneficial multiplier effects within the national economy.

However if that original £1 million is all spent within one local urban area it does not follow that the area will experience anything like the same multiplier effect. Even taking the 50 per cent spent upon direct employment of construction labour, it does not even follow that these will be local people who will spend their incomes within the local economy. Some workers, and especially the higher paid skilled workers and professionals might live many miles away and simply devote part of their time to this contract. The 10 per cent spent on imports is lost anyway, but neither is there any guarantee that any of the remaining 40 per cent of expenditure on raw materials, etc. will be spent in the local economy. All we can be sure of is that it will be spent within the national economy.

This has led some commentators, such as Jacobs (1984), to call for cities to undertake strategies of 'import-substitution' in order to increase the multiplier benefits of expenditure within their own boundaries. This kind of approach has been tried by a number of British local authorities who have attempted to oblige suppliers and contractors to use locally produced materials and seek local sub-contractors when undertaking work for the authority. At best such strategies have been only marginally beneficial to these cities. The problem with such strategies is that they distort the market, prevent that expenditure from occurring in some other (possibly equally depressed) location, and possibly have an adverse effect on total national levels of productivity and a detrimental effect on national economic growth.

For this kind of reason much of the conventional wisdom of urban economic regeneration has turned to two other approaches to the problem:

the first is to cross-subsidise between prosperous parts of the country and depressed areas; and the second is to search for new indigenous growth within depressed local economies through increasing the competitiveness of local firms and the birth of new small firms.

The first alternative is mainly achieved through the workings of the welfare state (the National Health Service, pensions and social security payments); national wage bargaining; and the Rate Support Grant, the mechanism through which the government balances the resources available to local authorities. While it was, for a short time after 1976, part of the Labour government's strategy to increase the proportion of Rate Support Grant going to depressed inner city areas, the incoming Conservative government was quick to reverse this trend in the 1980s leading to an inevitable negative effect upon these local economies. As Robson has commented:

> In London, boroughs designated under the Urban Programme received some £300 million between 1979/80 and 1983/84, but they 'lost' five times that amount (£1,530 million) through reduction in Rate Support Grant (Lever and Moore, 1986, p. 145).
>
> (Robson, 1988, p. 98)

It has also been argued that when the locational impact of other sectors of government expenditure are taken into account, such as defence, higher education, and the civil service, then the balance is very much against the economic regeneration of inner cities.

The second alternative has been developed more vigorously by this government through a series of policies and programmes designed to reduce industry's costs within the inner city and in offering advice and several forms of financial assistance to new and expanding small and medium sized firms.

3.8 Conclusions

The processes of urban renewal result from the interplay of economic forces whose nature and power must be recognised in any interventionist policies, whatever their motives. The demand for commercial and industrial construction is derived from the needs of capitalists for buildings in which to carry on their activities. There may be some speculative building ahead of demand and there may be situations in which suppliers can modify or stimulate demand. Historically the state, through regional policy and more recently through urban regeneration policies has had some effect in re-locating demand. But these are marginal changes and ultimately the nature, scale and location of demand depend upon local costs and trends in final demand for goods and services. Thus a major influence of the state on the location and scale of building and urban renewal activity is not through regional policy or urban regeneration policies in any narrow sense but through its influence as a consumer: as the monopsony purchaser of most investment in civil engineering and construction for the defence, health, education and other welfare services.

The proportion of construction activity that occurs as urban renewal rather that new building on greenfield sites is increasing. Partly this is an inevitable outcome of an ever growing and aging built stock continually requiring refurbishment, conversion or replacement to market demands. But also there is a degree to which this shift towards renewal has been encouraged through changes in government policies: tighter restrictions on peripheral development; enhanced subsidies and managerial changes to promote the redevelopment or refurbishment of obsolete urban fabric.

As society becomes increasingly dependent upon the maintenance of property values for its financial security it seems probable that the mainstream property market will become even more focussed upon standardised tradeable commodities with the implication that only conventional developments in favoured locations will attract institutional funding.

4 Social Aspects of Urban Renewal

The essential characteristic of urban renewal is that it brings about change in the use or occupancy of urban land and buildings and therefore results in changes in where, how and under what conditions people live. For some people these changes bring about improvements in living conditions, for others things get worse. This chapter presents a brief discussion of (i) the nature of urban population trends and their implications for urban renewal and (ii) consideration of the social needs of certain vulnerable groups within the population likely to be affected by urban renewal activity. Thus this is not a comprehensive account of the consequences of inner city living for the poor, disadvantaged and vulnerable but is confined simply to the major direct effects of urban renewal and its social impact.

4.1 Population

It is well known that the process of urban renewal, including decentralisation and dispersal, has led to a dramatic decline in the population of inner city areas. Population decline has been a feature of most British and North American conurbations since the 1950s and in continental Europe since the late sixties (Hall and Hay,1980). The two basic trends have been referred to earlier: decentralisation, whereby the conurbation spreads outwards to cover more land while fulfilling broadly the same economic functions thereby lowering average densities throughout the conurbation but particularly in the inner core areas; and secondly the absolute decline in the size of some conurbations as population and employment sources move out of the conurbation completely to other regions.

Figure 4.1 shows how all the major English conurbations have experienced population dispersal in recent years while most have also experienced absolute decline.

Insights into the detailed nature and social consequences of these trends can be developed through analysis of the situation in an individual conurbation. Here the case of Liverpool is examined. Although the city has suffered particularly severely in recent years, the trends found are

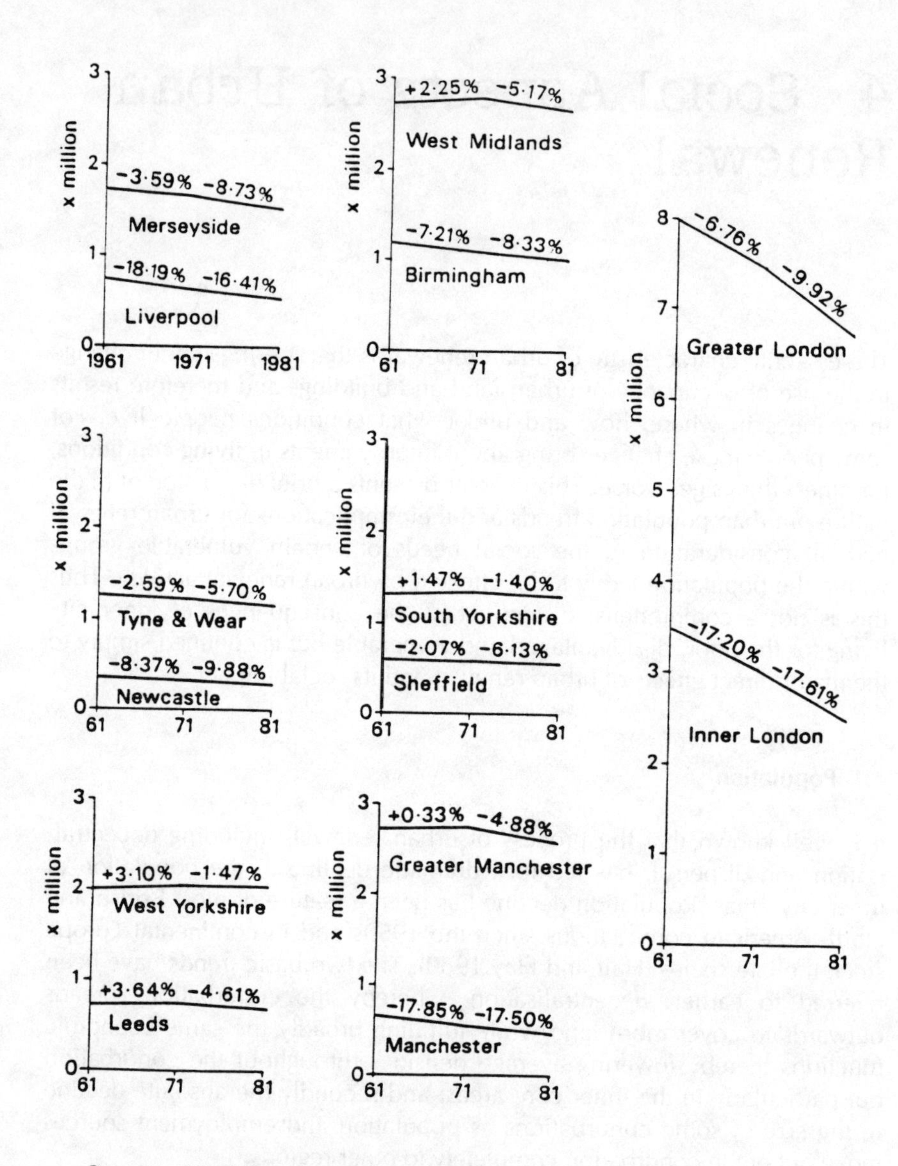

Figure 4.1 Population decline and dispersal in major British conurbations

replicated to a greater or lesser degree in most other conurbations and major cities, except London where more complex population movements are taking place.

Much of the decentralisation of population that occurred during the post-war period was brought about as a result of housing policy. On the one hand, the public housing sector dealt with the problem of obsolete inner area housing by demolition, replacement at lower densities, and overspill in peripheral council estates and new towns. On the other hand the private sector was permitted to build large amounts of new owner occupied housing on greenfield sites at ever increasing distances from the city.

Both these circumstances have now changed. The building of peripheral council estates has virtually ceased. This is partly due to some reduction in housing need pressures, partly government policy no longer favouring public sector housing, and partly as urban renewal policy has shifted away from clearance and replacement towards mass housing improvement. Within Merseyside (Greater Liverpool) and some other conurbations there has also been something of a reduction in peripheral private housing development. Similarly, this too is due to slackening demand pressures (and more profitable housing investment opportunities elsewhere in the country), but has also been influenced by much tougher Green Belt and agricultural land protection policies and the opening up of opportunities for private housing investment in the inner city. In Merseyside, as a result of these changes, there has been a slackening in the rate of population loss from the inner core of the conurbation, falling from a high rate of loss of 1.6 per cent of the population per year in the 1970s to less than 1 per cent per year as we come into the 1990s.

This population decline does not occur in a smooth, evenly balanced way. Rates of population change vary considerably across different parts of the city, depending upon local economic conditions, the spatial impact of housing policy and natural change (birth and death rates) within a local population.

Neither does population decline evenly through age groups or socio-economic groups. It is the younger people who are more likely to migrate, both to the periphery of the conurbation (for housing reasons) or to other regions (for employment and other reasons). Thus, the elderly tend to be left behind in the migration process. Figure 4.2 shows how Liverpool has suffered particularly from the loss of young people, especially school-children (with inevitable consequences for the provision and efficiency of inner city schools). This is due both to natural change (falling birth rates) and out-migration of young families. People established in a house and economically active (in the 15–59/64 age group) and the elderly have a greater tendency to remain in their existing homes. It will also be apparent that the over-75s were the only age group to have increased in both absolute numbers and as a proportion of total population.

It is not only the elderly who tend to be left behind in the inner core of the conurbation: it is also the poor and the less skilled. Thus, Liverpool not only has fewer people in the top status groups (employers, managers and professionals) and more people in the lower groups (unskilled and semi-skilled) than the national average, but it is gaining high status groups at a considerably slower rate, i.e. Liverpool is diverging from the national trend (see Figure 4.3).

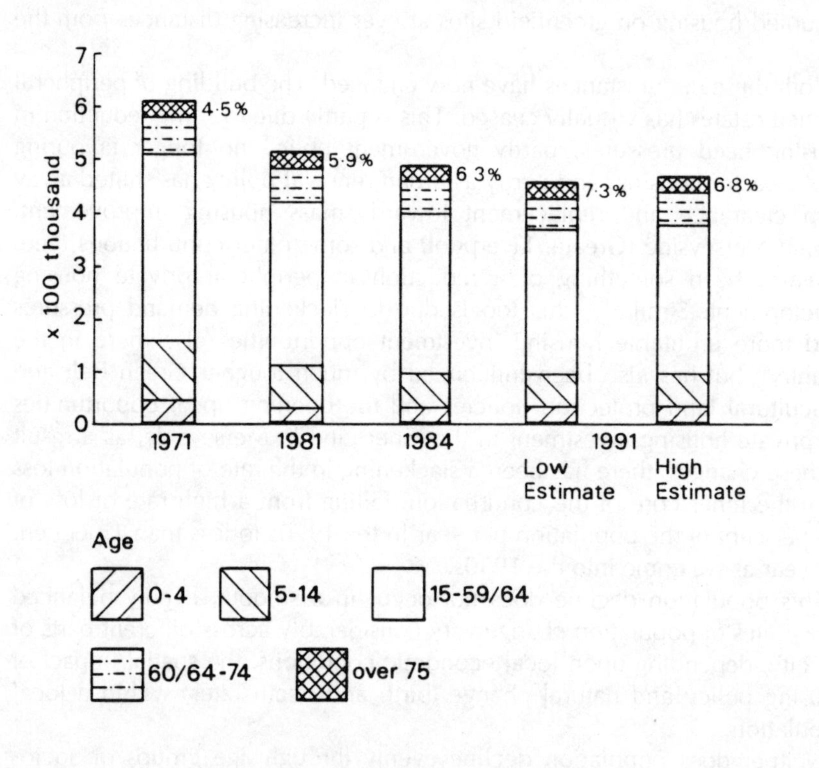

Figure 4.2 Population trends in liverpool

Equally important are the effects of unemployment and job loss. Between 1971 and 1981 Liverpool lost 38 per cent of all its manufacturing jobs and 17 per cent of its service sector jobs. This compares with a national trend of 23 per cent reduction in manufacturing jobs and 11 per cent *increase* in service employment. The recent trends in unemployment show that the

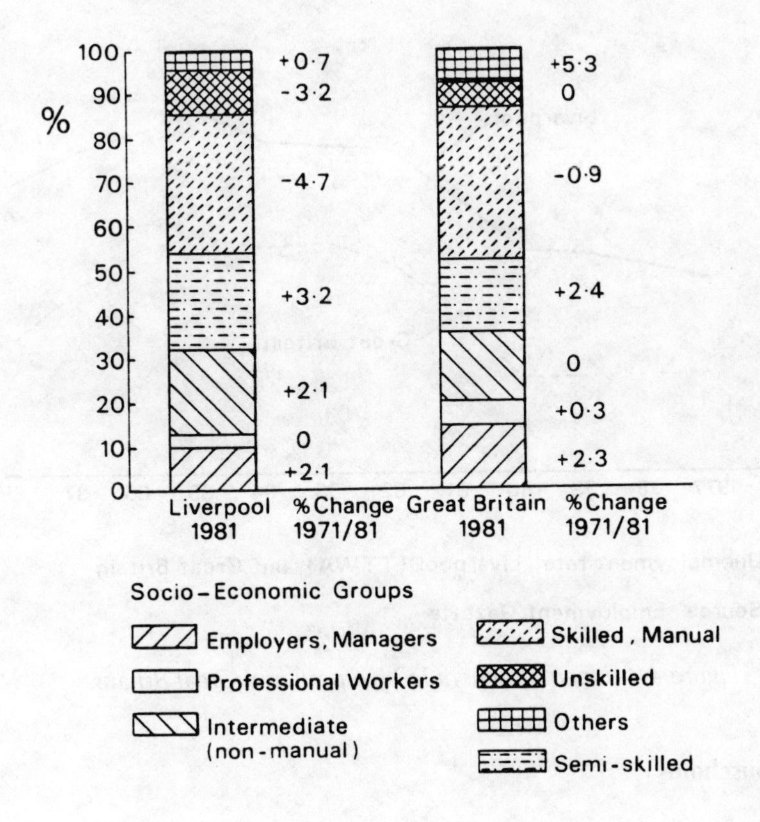

Figure 4.3 Changes in socio-economic structure: Liverpool and Great Britain

inner core of the Liverpool conurbation has continued to have an unemployment rate well above the conurbation average and nearly double the national average (see Figure 4.4).

The combined impact of all these changes in population structure is such that as time goes on, the remaining population is increasingly less able to provide for its own needs for age and skill reasons and is increasingly less able to find the income, through employment, necessary to enable effective competition in the housing market. Thus, the per capita incomes of those in employment have to be stretched ever further to meet the ever growing needs of an increasingly dependent population.

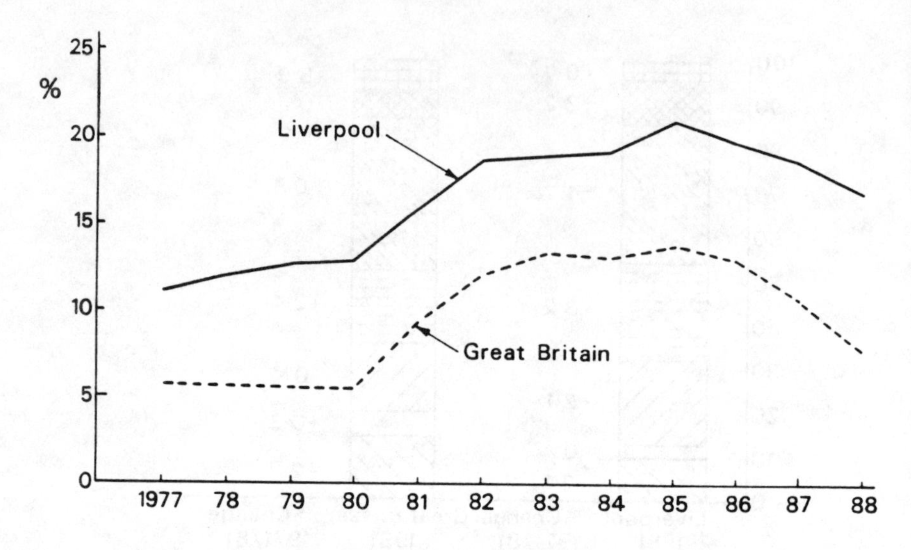

Unemployment rate, Liverpool (TTWA) and Great Britain

Source: Employment Gazette

Figure 4.4 Unemployment in Liverpool and Great Britain

4.2 Households

In considering the effects of urban renewal on housing needs it is not sufficient to analyse population trends. A significant influence is the way in which the population then converts itself into households; for it is households, rather than individuals, that determine housing requirements.

For many years, until the end of the 1960s, Liverpool continued to see a rise in the number of households, despite a falling population. This was because until that time the effect of a declining average household size (tending to create more households) was greater than the effect of population decline (tending to create fewer households).

Between 1971 and 1981 average household size fell from 3.09 to only 2.59 persons per household. Two factors underlie this trend. The first was the decline in family size: both through families having fewer children and through the rising number of single parent families. The second was the increase in the numbers of small, childless households. This group of households includes a growing number of young adults who choose to live away from the parental home for educational, employment or social reasons; a growing number of two person households, such as couples who have postponed or foregone having children; and a growing proportion of the elderly who live alone or separate from their extended family. The effect on household structure can be seen in Table 4.1.

Table 4.1 Household structure in Liverpool

	1971	*1981*
Households	194,399	180,507
Average household size	3.09pph	2.59pph
Single person households over pensionable age	12.1%	15.7%
Single person households under pensionable age	7.8%	9.4%
Two person households	27.8%	28.3%
Households with children	52.3%	46.6%

Source: Census of Population

The consequence of these trends is an exacerbation of the dependency problems referred to above.

Today the declining and increasingly poor 'support' population of Liverpool, and other similar conurbations, has to provide an increasing number of dwelling units per capita in order to house itself. It should not then be surprising to find that these inner city populations are increasingly taxed (through local rates) and increasingly reliant upon outside financial support systems (through government subsidy) simply in order to maintain the same housing conditions and levels of local services that existed a decade ago. There is no doubt that any failure to increase these external financial supports towards inner city areas in proportion with growing needs inevitably leads to a downward spiral in living conditions.

4.3 Communities and Neighbourhoods

The idea that residential areas within cities contain socially cohesive groups of people, or communities, has been with us for many years. Many sociological studies have analysed the nature of these communities both within inner city working class neighbourhoods (Young and Willmott, 1957; Gans, 1962 for example) and suburban middle class areas (Gans,

1967 for example). It is commonly suggested that the inherent stability (i.e. lack of spatial, social or occupational mobility of residents) of many inner city working class areas leads to the proliferation of kinship ties within the area and the establishment of long term friendships with peers. The suburban middle classes, it is suggested, are more transient, develop much shorter term local friendships, but have the necessity and the means to retain long-term kinship and peer group relationships across great spatial distances.

Thus it has been argued that urban renewal, which tends particularly to affect inner city working class areas, has damaging social effects on these communities and psychological effects upon individuals. The suburban middle classes are less often affected by urban renewal and are more able to cope with the enforced changes because of their more limited local dependence, their greater affluence and social skills.

However the situation in inner city areas is more complicated than this. The development of these working class communities depends upon the long term stability referred to above. By no means all inner city residential areas are of this nature. There is a world of difference between the late Victorian terraced family housing areas of Anfield in Liverpool or, say, much of Newham in London, and the transient populations to be found in the rooming-house areas of Toxteth or Earls Court. Furthermore the existence of close-knit communities is only one aspect of inner city working class life. The problems of high density living, poor quality housing with inadequate amenities, and social conflict are other common characteristics of these neighbourhoods. Thus for residents who are moved out of an area there is a trade-off between these various costs and benefits. For many the most acute consequence of moving is a rise in the costs of housing and journeys to work. Nevertheless Knox summarises the work of a number of researchers, saying that:

> It is clear that resettling working-class families in suburban estates does result in some disruption of primary social ties with the result that, for a time at least, social cohesion is reduced ... An important contributory factor in this respect is the aloofness generated by status uncertainty as the result of movement to a new and socially unknown environment ... This, together with the reduction in frequency of contact with kinfolk, leads to the development of a more home-centred way of life, so that many suburban public housing estates are characterized by low levels of neighbouring and a lack of social participation in organized activities in clubs and societies.
>
> (Knox, 1982, p. 47)

For the individual the consequences of a forced move depend upon a host of factors ranging from personal psychology to age, occupation, income and indeed, the attitudes and actions of local bureaucracies. For children it may mean a new school, new teachers, a disrupted curriculum, and making new friends. For a working adult many of the pressures above will apply. For an elderly person these social problems may be felt particularly severely. These were some of the arguments used in the late sixties/early seventies to turn housing renewal policy away from slum clearance and overspill towards area housing improvement. However, it should not be assumed that housing improvement is without its personal impacts and social costs. Although the household may benefit from no change in its surroundings or mobility it is likely to have to pay a significant proportion of the costs of improvement (amounting to several thousand pounds if an owner occupier) or pay substantially increased rents (if a tenant). Owner occupiers who improve their property frequently have to pass a number of months 'camping out' in different rooms as work progresses. Tenants are more likely to be decanted to another property while the work is carried out and moved back on completion. Neither situation is free from financial cost, inconvenience or personal stress.

4.4 Women

Another useful dimension to the analysis of urban renewal and its consequences is the consideration of the implications of change for women. Not only does such analysis inform us about the relationships between women and the urban environment but it also gives more general insights into the impact of changes on family life, on children, on personal safety, and on personal opportunities.

Writing in 1984 about the implications for women of the recent revitalisation of many North American inner cities, Holcomb contrasts two situations: the affluent 'career' woman (especially if childless) and the non-affluent woman. In the first case:

> Gentrified neighbourhoods ... are typically architecturally distinctive (with) dwellings which were originally large single family homes, row townhouses or even industrial and commercial buildings. As their exteriors are restored to some semblance of their former glory, with masonry detail and pastel paint, their interiors are modernised with well equipped kitchens and sybaritic bathrooms ... These homes are set in pleasant pedestrian neighbourhoods with brick sidewalks, generous plantings and attractive views. There is usually easy access to stores stocked with a variety of upmarket consumer essentials (French roast coffee beans and potable white wine) and providing such services as dry cleaning and catering. Such environments, if not perfect, are certainly pleasant residences for both women and men of means. While lacking the community kitchens and kitchenless homes of the feminist utopia ... kitchens equipped with food processors and microwave ovens make

> culinary art a recreational possibility, while the abundance of nearby take-out or eat-in food establishments reduce the need to cook at home. The typical proximity of gentrifying neighbourhoods to those with low-income residents facilitates access to paid domestic labour ... The professional woman lives close to employment opportunities, commuting is minimised for both sexes, and art galleries, theatres, opera and health spas are easily accessible. In brief, if a woman has money and does not have children, life in a revitalised American city can be pleasant.
>
> (Holcomb, 1984, p. 20)

On the other hand:

> Unfortunately, most of the jobs created by redevelopment which are likely to be available to women are also low paid and lack much potential for advancement ... Redeveloping the physical fabric of a city does not necessarily change its occupational segregation. While there are undoubtedly women who benefit from the economic restructuring of cities, many are gaining little ... As consumers, non-affluent women are further penalized when specialized (and costly) goods and services proliferate, often at the expense of lower order necessities. Delicatessens and gourmet ice cream parlours replace food supermarkets, used furniture stores are transformed into antiques emporia, and running shoes are sold where the cobbler once replaced worn soles. The diversion of public monies from social services to infrastructure improvements, downtown police protection, and tax abatements for corporate headquarters, reduces the range and availability of such services as subsidised day care, meals on wheels, or pre-natal health care – the consumers of such services being disproportionately female.
>
> (Holcomb, 1984, p. 22)

While these extracts are drawn from United States experience their British parallels, albeit in a muted form, are obvious.

Much of the British government's 'streamlining' of the country's social economy has particularly affected women. The reductions in public transport subsidies reduce the mobility of those without access to a car (mainly women); the introduction of 'market' pricing to many areas of public service, such as staff crèches in NHS hospitals, reduces the employment possibilities and incomes of women; while reductions in local authority social services and the increasing trends towards 'community care' for the mentally sick and elderly actually does little more than place acute physical and psychological burdens on individual women. Indeed it has been suggested that for a local neighbourhood to be supportive of the needs of elderly women, single parents and working mothers there needs to be a range of good quality housing options, social and community services and public transport. (Wekerle,1988, p. 17). All three of these supports have been undermined by government action during the 1980s.

Nevertheless, there have been many improvements to the urban physical environment in recent years that could be regarded as beneficial to women: better street lighting improving personal safety; better provision for prams to be manoeuvred up and down slopes, across road junctions and onto public transport; more pedestrianisation of shopping streets, and so forth. But two comments have to be made. Firstly much of this provision is 'too little too late'. Good secure environments are very much the exception in the existing environment and by no means universal in newly designed or redeveloped areas. The second point is that many of these improvements, and especially the way the decisions are made, are patronising and often no more than token gestures (usually by men) based on stereotyped perceptions of womens' needs and desires.

In an effort to overcome some of these problems a number of professional bodies involved with different aspects of urban renewal, notably the Institute of Housing, the Royal Institution of British Architects and the Royal Town Planning Institute (RTPI) have set up working parties or produced reports on these issues.

Ideas favoured in a report to the RTPI mainly focus upon the aim of attracting more women into the profession, both for its own sake and to help develop a stronger feminist analysis of town planning and urban renewal issues and policies. The proposals include: rewriting careers information and promotional literature to appeal more to women; changing the content and structure of town planning education to reflect the interests and needs of women students; appointing officers to the professional body with specific responsibility for women's matters; establishing contact groups and other means of raising women's confidence within the profession. They also include encouraging employers in the provision of crèches, flexible working arrangements, job sharing and reducing the barriers to a return to work after childbearing; and, perhaps above all, encouraging male town planners to acknowledge the existence of sexism as a problem and the legitimacy of feminist analysis.

4.5 Race

According to the 1981 Census around 2.2 million people live in households where the head of household was born in the New Commonwealth or Pakistan. This is undoubtedly a substantial underestimate of the present black population of Britain since the Census is now quite out of date and excludes black heads of households born in this country (a rising proportion). This population has certain characteristics and needs that require distinct responses from the state. The majority of this black population is concentrated in the inner cities and is therefore disproportionately likely to be vulnerable to or affected by urban renewal processes compared with the population as a whole. For these reasons it is important that those concerned with intervention in urban renewal should be particularly aware of the racial aspects and implications of policy.

Until recently the black population tended to have a younger age profile than the national average, placing greater demands upon education, child welfare and youth services but correspondingly less demands on the health system and services for the elderly. The economic activity rate for black males of working age is estimated at 93 per cent compared with 91 per cent for the white population (Smith, 1976) but the black population is less likely to be employed in high status professional or managerial occupations and far more vulnerable than the white population to unemployment. The housing conditions of the black population are also likely to be worse than average. These points are illustrated in Figure 4.5.

According to a joint working party of the Royal Town Planning Institute and the Commission for Racial Equality:

> All housing policies will have a disproportionate effect on some section of the the black communities; those relating to council housing will particularly affect West Indians while those concerning housing renewals and improvement in the inner city will have important consequences for Asians. In formulating development plan policies and proposals, local planning authorities have a duty ... to survey the size, composition and distribution of the population of their area and thus are properly able to tailor land-use policies and proposals (and hence to some extent development control) to meet any special needs arising from the presence of groups in any locality.
>
> (RTPI/CRE, 1983, p. 32)

Recent immigrants are likely to be subject to major changes in their social conditions and surroundings and vulnerable to stress in many forms. Living in an area undergoing urban renewal compounds and exacerbates this stress. Many Asian immigrants, particularly women, do not speak English as a first language and have particular communication needs in the realm of public participation in urban renewal policy making. The RTPI/CRE report suggests that translation and interpretation facilities should be available as a matter of course in multi-racial areas and that the employment of bilingual staff should be encouraged (RTPI/CRE,1983, p. 33).

There are cultural differences between races that should be recognised and accommodated through urban renewal policies. For example some groups have a greater preference for living in larger extended family groups than is normal in this country. Different religious groups have different needs both in terms of buildings and the range of religious and social activities that take place within them.

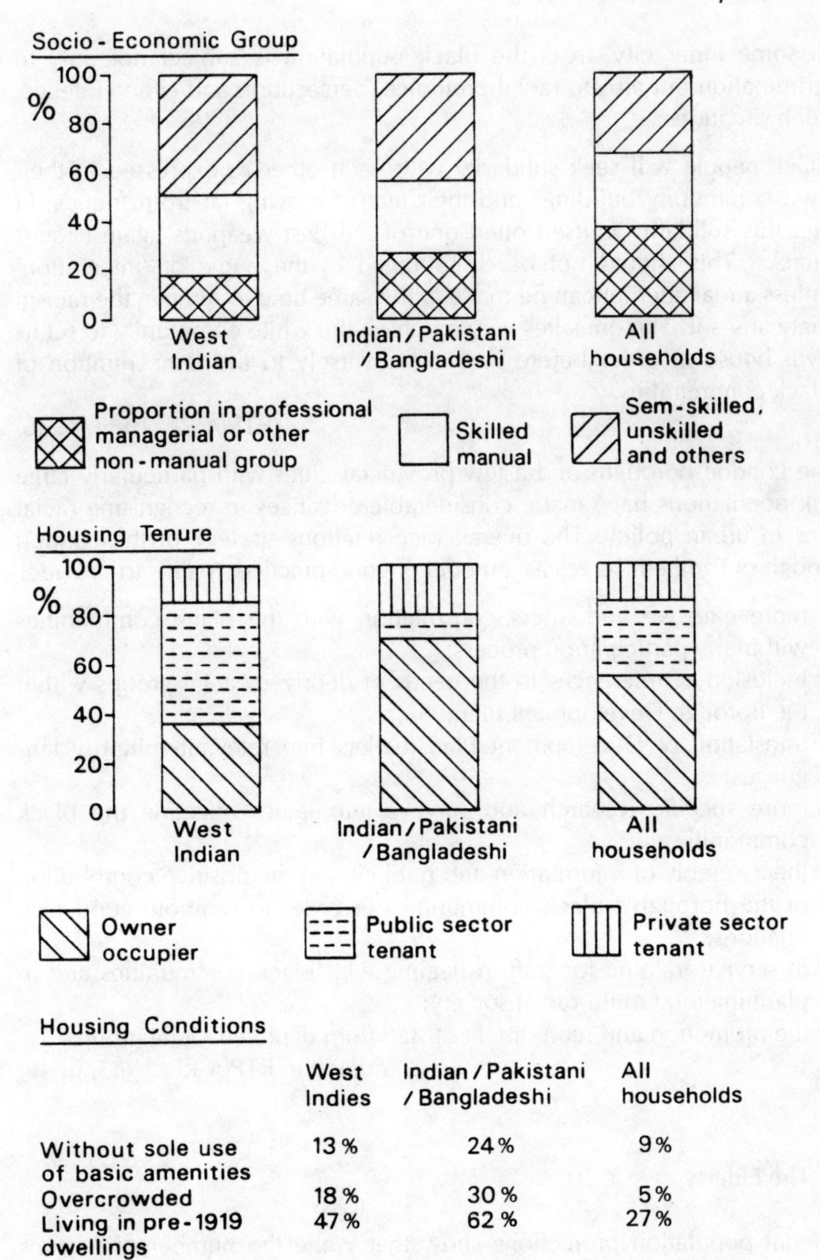

Housing Conditions

	West Indies	Indian / Pakistani / Bangladeshi	All households
Without sole use of basic amenities	13 %	24 %	9 %
Overcrowded	18 %	30 %	5 %
Living in pre - 1919 dwellings	47 %	62 %	27 %

Source: National Dwelling and Household Survey, HMSO 1979

Figure 4.5 Some economic and social conditions of the black population

In some inner city areas the black population is subject not only to discrimination but also to racial prejudice, persecution and even violence. In such situations:

> Black people will seek solidarity with each other as expressed in their own community buildings and their fear of moving far from friends. In fact this solidarity is itself often one of the best weapons against racist attacks. This should not be discouraged in the name of 'integration' unless a real attempt can be made at the same time to remove the racism that causes it. The onus lies very much on the white community to set its own house in order before reacting adversely to the concentration of black communities.
>
> (RTPI/CRE, 1983, p. 36)

Some London boroughs and a few provincial cities with particularly large black populations have made considerable advances in recognising racial issues in urban policy. The overall race relations strategy of the London Borough of Lambeth serves as a model of good practice. It aims to provide:

- more effective and direct consultation with the black communities within the participation process;
- inclusion of references to the needs of deprived racial groups within the Borough Development Plan;
- translation of Development Plan leaflets into relevant minority languages;
- more specific research and surveys into issues affecting the black communities;
- the assembly of information and publicity on the positive contribution of the Borough's black communities in order to promote good race relations;
- in-service training for staff in dealing with black communities and in planning for a multi-racial society;
- the promotion and recruitment of staff from deprived racial groups.

(quoted in RTPI/CRE, 1983, p. 46)

4.6 The Elderly

National population projections show that while the number of persons over pensionable age will increase only slightly over the next twenty years, the numbers of the very elderly (75 and over) will rise to 3.7 million (about 20 per cent higher than the present figure). For urban renewal policies, especially in housing provision these are significant trends.

With regard to general housing provision for the elderly the Department of the Environment has issued the following advice:

> This group (the very elderly) are more likely to have difficulty in coping with their existing accommodation and to require additional care and support ... Special housing provision for elderly people who have difficulty in managing their existing accommodation may take a number of forms:
>
> a. small, mainstream dwellings;
>
> b. small dwellings designed for the elderly including features such as whole house heating, raised electric sockets, special bathroom fittings, improved accessibility within and to the dwelling, etc;
>
> c. sheltered housing with a warden service and an alarm system, and optional community facilities.
>
> In addition, full-time residential accommodation or hospital care may be required, while domiciliary services add a further dimension.
>
> (DOE, 1980, p. 75)

We can characterise the elderly into three distinct groups as far as urban renewal processes are concerned. Firstly, there is a rapidly growing number of relatively affluent pensioners living on substantial private pensions usually in owner occupied housing which they own outright. The spending power of this group expanded so much during the 1980s that it supported a major growth sector in the speculative housing market: secure, easily maintained, owner occupied semi-sheltered accommodation for the elderly. This market has been particularly attractive to speculative developers in urban renewal situations where the redevelopment of small plots to provide new accommodation and the conversion of larger older buildings into small flats provides a lucrative investment.

The second group are generally less affluent but have managed to enter the shelter of the state welfare system, living in council or housing association accommodation and supported by various housing and social security subsidies and support systems. If they are affected by urban renewal the state will normally, at least at a minimum level, take care of their needs.

The third group are the poor elderly who are outside or only partially covered by the state welfare system. They may be poor owner occupiers or tenants living in property they do not have the means or the ability to maintain. Many of these people form the residual population of inner city areas, left behind by the outward migration of the young and the skilled, and they are particularly vulnerable to the consequences of urban renewal processes.

4.7 Conclusions

At the macro scale, between urban areas, the dynamics of population movement are primarily determined by the shifting labour requirements of industry and commerce. The only major exception to this is in patterns of retirement, where environmental considerations are important. These movements lead to differences between urban areas in terms of population structure: age profile; socio-economic or occupational composition of the population; and, more marginally, household structure, and the gender, cultural and racial composition of the population. This resulting social structure is a significant influence upon the nature of the urban renewal processes that might take place in the area, and is also indicative of the probable robustness or vulnerability of that population to withstand, manipulate, or be disadvantaged by the impacts of urban renewal. Within urban areas economic forces, particularly personal incomes and status, are still important influences upon population movement and settlement patterns but here the impact of local housing, planning and transportation policies become significant.

Because those social groups most vulnerable to change (low income groups, racial minorities and the poor elderly) frequently live in precisely those parts of the city most likely to be affected by urban renewal processes it is important that policy makers understand the needs and aspirations of these groups. In these situations detailed local sociological research, awareness training for urban renewal staff, community participation and the devolution of decision making become important elements of good social policy.

5 Management and Organisational Theory for Urban Renewal

The organisation and management of intervention in urban renewal is a complex activity involving a number of agencies each with its own, often conflicting, objectives; operating in different modes and at different speeds. In reality little attempt is made to coordinate these various activities beyond (i) placing considerable faith in market mechanisms, and (ii) relying on the disjointed incrementalism so aptly described by Lindblom in the 1950s (1959). That is to say that each new policy or decision is a reaction to some perceived problem, usually takes the form of an incremental adjustment to existing policy and is based upon a limited consideration of a small number of well tried alternative solutions. The resulting change then leads to further incremental adjustments in policy. However, to the extent that some agencies do wish to improve their performance in urban renewal and do wish to manage these processes, there are some lessons to be learned from management and organisational theory. Much of this literature has been written in the context of managing a private business, nevertheless many of the concepts and ideas transfer suprisingly well both to the management of the public sector agencies involved in urban renewal, and to the management of urban renewal programmes and projects.

5.1 Planning and Decision-making Processes

Much public intervention in urban renewal is concerned with the imposition of a planning process to bring order and control to the market. The simplest view of the planning process can be described in the rational model (see Figore 5.1). This is a model that has its origins in management literature but is also prominant in urban planning theory (see for example Hall, 1975).

While this model shows the process to be cyclical, it is conventional to describe the process by breaking in at the top of the circle: problem identification.

Problem identification and objective setting have to be taken together for there can be no problem without an objective. For example, unemployment only becomes a problem when full employment has been defined as an objective, unfit housing only becomes a problem when

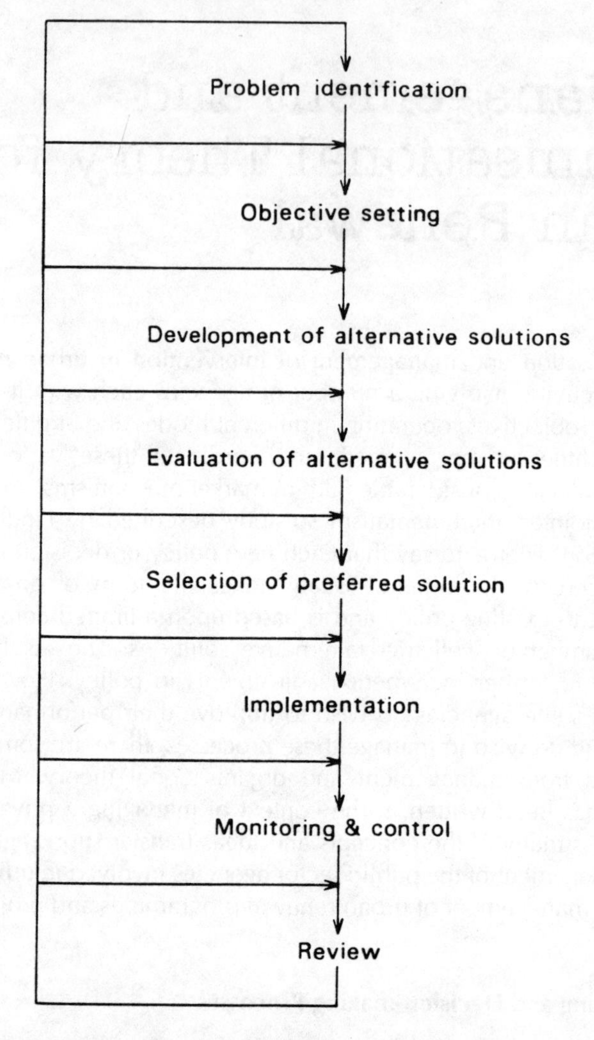

Figure 5.1 The rational planning process

fitness standards have been established. Objectives are defined by different sections of government, by other public agencies, by private organisations and by individuals. Their definition is a political process and their relative importance and weight shifts over time and space as social and economic conditions and pressures change.

In order to move towards these objectives various potential solutions may be considered. In a truly comprehensive rational model all possible solutions would be on the agenda but in practice only a small number of marginal or well tried changes in policy are normally considered. These alternatives may be tested or evaluated against three criteria: effectiveness, efficiency and equity. The effectiveness criterion is concerned with assessing the degree to which each potential solution achieves the stated objective; this in turn implies that objectives should be specified in a manner that is capable of measurement, such as to halve unemployment within two years, or to provide all dwellings in the area with all standard amenities within five years. The efficiency criterion is concerned with the comparison of alternatives in terms of their relative costs: the ratio of outputs to inputs; this in turn raises a series of questions about the choice and measurement of outputs and inputs. In Figure 5.2 the example of advance factory building is used to illustrate a range of possible different interpretations of efficiency and effectiveness.

The equity criterion is concerned with the distributional effects of policy: the questions of who pays and who benefits? Which categories or groups of people will be made worse off by the proposals and which categories or groups will be made better off? The problem of measuring the economic costs and benefits is a major issue in economic and management theory. Fundamentally the questions that have to be faced are: what costs and benefits are to be included in the calculation? How are they to be valued or costed, particularly when (as with many environmental costs) there is no immediate money value that can be attached? How are costs and benefits occurring at different points in time going to be brought to constant prices (i.e. what discounting rate is to be used)? Finally the evaluation process will be concerned with assessing the flexibility (ability to accommodate changed circumstances) and robustness (ability to withstand change) of each alternative.

In the logical sequence of events the selection of a preferred alternative follows the evaluation process. However, it must be remembered that the various elements of evaluation cannot be 'added up' to provide a sum which is an answer. There may be a number of objectives, each of which is achieved to a different degree by each policy alternative; the policy that is most effective may not be the most efficient; each policy alternative may have different distributional consequences and the importance attached to each outcome will depend upon organisational priorities. Thus the selection process involves choices based upon value judgements, in other words it is a political process in which formal evaluation merely provides an input of useful information.

In the implementation of the preferred policy four elements of organisational context have to be considered in a further refinement of the evaluation process.

(i) Does the organisation have the legal power to carry out this task?
(ii) Does it have the necessary resources (whether measured as money, or real economic resources of land, labour and capital)?
(iii) Is there sufficient political and community support?

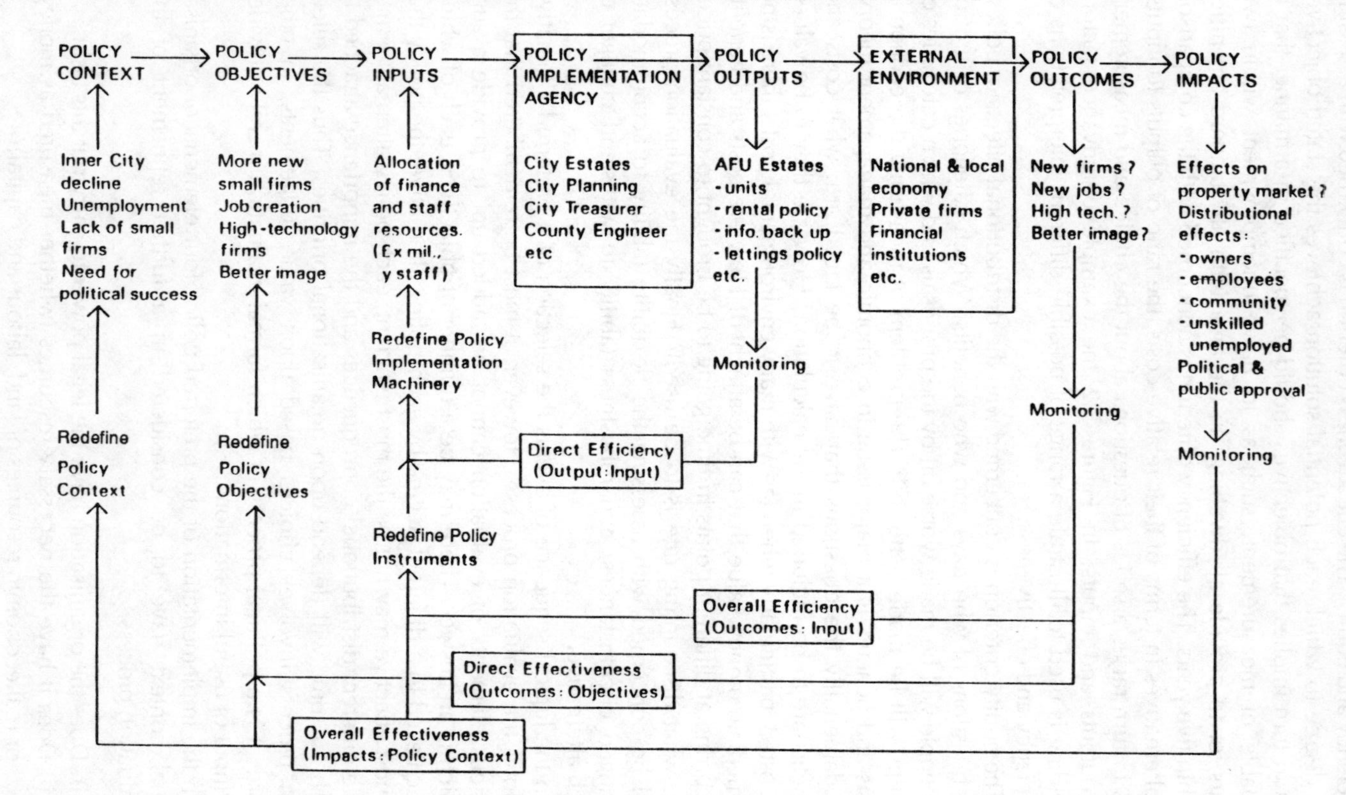

Source: After J Herson, Liverpool Polytechnic, 1983 originally derived from D336 Policies, People & Administration Open University 1981.

Figure 5.2 The evaluation of efficiency and effectiveness criteria

(iv) Is there an appropriate organisational structure and capacity to undertake the task?

Once agreed, the implementation decision has to be executed, enforced and controlled. Monitoring has to take place both to ensure that the policy is being implemented as intended and to discover its effectiveness. At any stage, and particularly if there is some evident failure of policy, the whole process can be reviewed. This may reveal that failure might be attributable to error at any stage of the process: inadequate problem identification, wrongly conceived objectives, erroneous evaluation, wrong choices or inadequate selection procedures, poor implementation or control procedures. One of the very important lessons from the history of urban renewal is that it is a long way from policy inception to final implementation and that much which appears to be bad policy may well turn out to have been bad implementation of good policy. For example much of the criticism of post-war slum clearance policy could be directed towards the implementation process rather than the original policy; likewise it could be argued that many of the failings of the Statutory Development Plan system could be attributed to poor implementation rather than to the quality of the original thinking and conception.

Simon (1960), writing in the early 1960s, was working with decision-making models very similar to that just discussed but he also distinguished programmed, routine, repetitive decisions from non-programmed, unusual, unexpected decisions. In his view the techniques of programmed decision making included habitual reaction, routine procedures, etc. and would equate with the incremental adaptation which Lindblom (1959) had suggested dominated most decision-making. Non-programmed decisions required processes of judgement, intuition and creativity: in other words a much higher level of intellectual response that in turn required higher levels of education and training.

A related classification of decisions is the distinction between the strategic and tactical levels. Strategic decisions are high level policy, long run, unprogrammed, directional and more concerned with relationships between the organisation and the external environment. Tactical decisions are low level, immediate, routine and concerned with the implementation of decisions taken at the strategic level.

Cyert and March (1963) saw organisations as being 'adaptively rational', responding and adapting their behaviour to a variety of internal and external constraints when arriving at decisions. They identified four features of the decision-making process which they claimed represented a realistic theory of decision-making:

(i) quasi resolution of conflict: complex problems were usually disaggregated into simple elements that could be resolved by individual parts of the organisation;

(ii) the avoidance of uncertainty: by limiting uncertainties in the external environment through improved knowledge or control;
(iii) problemistic search for alternative solutions: the search for solutions tended to be limited, conservative and simple minded;
(iv) organisational learning: the organisation learned from its experiences.

Thus on the one hand the normative theory represented by the rational planning process suggests an 'ideal' decision-making process and on the other hand positive theories (Lindblom, Cyert and March) attempt to explain the way decisions are actually made. The former is useful in the construction of logical sequences of events in the preparation of urban renewal policies and programmes while the latter assist with the understanding and interpretion of actual organisational behaviour in urban renewal decision making.

5.2 The Management Function in Urban Renewal

A number of writers have sought to describe and analyse the management function and the roles of managers within organisations. Again it is useful to consider this work and its transferability to the inter-organisational or multi-organisational environment of managing urban renewal.

One of the earliest classical writers on management theory was Henri Fayol. In Fayol's view (1949) management is seen as having five functions:

(i) The first is to forecast future changes in circumstances, opportunities or problems and planning organisational responses to accommodate, withstand or take advantage of such changes (i.e. to prepare a strategy).
(ii) The next requirement is for operational and tactical decisions to be made within the organisation to ensure that the desired strategy can be followed.
(iii) These decisions then have to be implemented (i.e. there is a command function in which management uses various means to 'get things done').
(iv) Responses from different parts of the organisation (or between organisations) have to be coordinated both in direction and in implementation.
(v) Finally these actions have to be controlled in order to maintain adherence to the chosen strategy.

There are obvious parallels with the rational planning process but most writers would now regard this as an over-simplistic explanation of the functions of management. In a more sophisticated analysis of the nature of managerial work (i.e. what managers actually do) Mintzberg (1973) has identified ten managerial activities contained within three role sets:

Interpersonal Roles	*Informational Roles*	*Decisional Roles*
Figurehead	Monitor	Entrepreneur
Leader	Disseminator	Resource allocator
Liaison	Spokesperson	Disturbance handler
		Negotiator

Although Mintzberg was writing principally in the context of the management of organisations it is possible to adjust the interpretation of these concepts and to apply the analysis to the management of the 'process' of urban renewal. Here the team to be managed might refer to a number of individuals or organisations coming together in a partial way for some period of time for some common activity or process, which could be any urban renewal programme or project.

Interpersonal roles

Figurehead All programmes and projects need a figurehead, someone who provides and indicates status and whom other organisations can identify as an entry point for communication.

Leader Leadership is a necessary component of organisational and process management. The leadership function is to state and clarify objectives, to build and maintain the project team and to facilitate work.

Liaison Being at the top of a hierarchy the manager is always one of the major, but not necessarily the only, focuses for liaison between team members. Closely allied with the figurehead role, the manager is also the key point in liaison with other organisations.

Informational roles

Monitor Monitoring progress in relation to stated objectives, resource availability and changing circumstances; the controlling of work processes, and the review of actions in the light of deviations from intended paths; these are all important management tasks.

Disseminator Receiving and passing on information from the external environment to appropriate team members.

Spokesperson The converse of the last role is the passing on of information from the process or project team to the external environment and linked to this is an advocacy or influencing role in relation to that environment.

Decisional roles

Entrepreneur Managers have to make decisions, both strategic and tactical, about the future of the programme or project, (i.e. the 'product' or the 'market'), and about the means of production or working arrangements.

Resource allocator Decisions have to be taken about the choice and quantity of resources to be employed and the allocation and prioritisation of resources to specific tasks.

Disturbance handler It is a manager's task to decide how to react to the unexpected.

Negotiator There is always a requirement for negotiation with others both within and outside the team about activities and resources.

One of the keys to success in urban renewal (as in any other managed activity) is that a management structure has to be devised and set in place that allows all and each of these management functions to take place. Management has to be given the appropriate power and authority to carry out these tasks.

5.3 Organisational Theory for Urban Renewal

Some useful insights into the nature and operation of organisations have been provided through the application of systems theory in the management field. A system, in the organisational context, may be defined as:

> an organised, unitary whole composed of two or more interdependent parts, components, or subsystems and delineated by identifiable boundaries from its environmental suprasystem
>
> (Kast and Rosenzweig, 1981, p. 98)

Any urban renewal agency is a system which can be represented in its environment as shown in Figure 5.3.

The 'goals and values' sub-system is a key element since the organisation must perform certain functions or meet certain goals in relation to society's needs in order to survive. The 'technical' sub-system represents the knowledge and technology available within the organisation. The 'psychosocial' sub-system refers to the individuals within the organisation: their behaviour, relationships, group dynamics and so forth. The 'structure' refers to the way in which the organisation is sub-divided (e.g. by task or by area) and the ways in which these sub-divisions are coordinated or integrated. The 'managerial' sub-system covers the whole organisation carrying out all the managerial functions and roles discussed above.

One of the characteristics of intervention in urban renewal is that it takes place in a relatively unstable and uncertain environment with few

a) The Organisation as an Open System

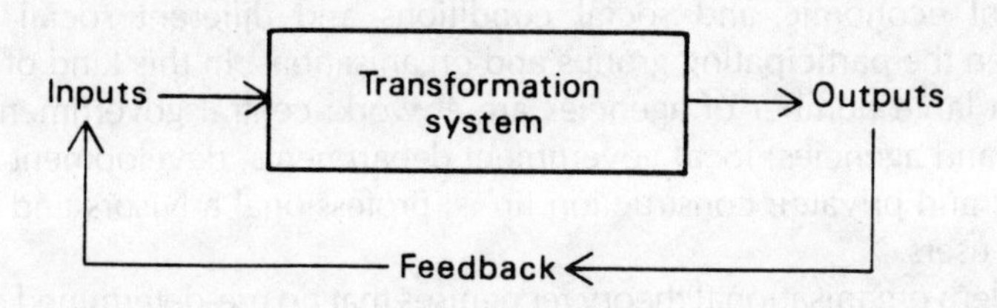

b) The Organisational System

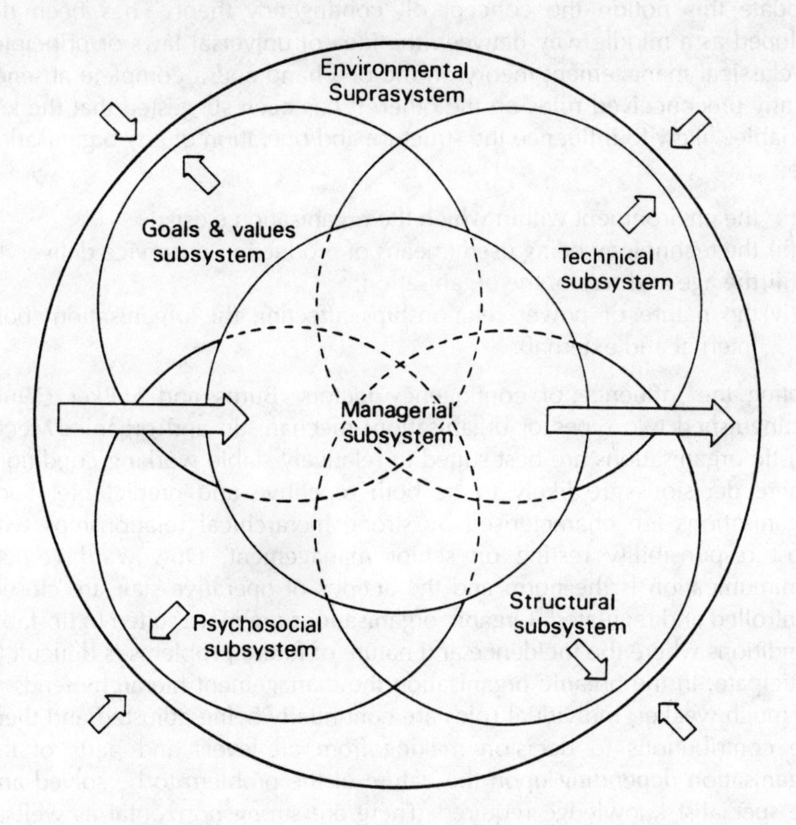

Input-output flow of materials, energy, and information

Source: after Kast FE & Rosenzweig JE, 1981, p109

Figure 5.3 The organisation as a system

repetitive decisions. Each investment decision takes place under slightly different economic and social conditions and different social relations between the participating groups and organisations. In this kind of environment a large number of agencies are at work: central government departments and agencies; local government departments; development agencies (public and private); construction firms; professional advisors and a variety of end users.

Modern organisational theory recognises that no pre-determined organisational structure can be prescribed as right for all circumstances but that organisational system design has to take account of the particular circumstances of each organisation, its function and environment. To accommodate this notion the concept of 'contingency theory' has been developed as a middle way between the idea of universal laws or principles of classical management theory on the one hand and a complete absence of any preconceived rules on the other. It has been suggested that the key variables likely to influence the structure and operation of any organisation are:

(i) the environment within which the organisation exists;
(ii) the technology being used (means of production or service delivery);
(iii) the age and size of the organisation;
(iv) the nature of power relationships affecting the organisation (both internal and external).

Noting the influence of contingency factors, Burns and Stalker (1966) distinguished two types of organisation: mechanistic and organic. Mechanistic organisations are best suited to relatively stable working conditions where decisions are likely to be both repetitive and predictable. Such organisations are characterised by strong hierarchical relationships with most responsibility resting on senior management. Downward vertical communication is the norm and the actions of operative staff are closely controlled and regulated. Organic organisations are more suited to unstable conditions where the incidence and nature of future problems is difficult to anticipate. In the organic organisation the management hierarchy tends to be much weaker; individual roles are continually being adjusted and there are contributions to decision making from all levels and parts of the organisation depending upon the nature of the problem to be solved and the specialist knowledge required. There are strong horizontal as well as vertical communication channels.

Lansley *et al.* (1977) applied a variation of Burns and Stalker's concept to the construction industry. They described a two-dimensional matrix with one axis ranging from high integration to low integration (i.e. the level of coordination, sharing of information, joint operation, etc.) and the other axis ranging from high control to low control (i.e. the level of managerial power and authority). This is shown in Figure 5.4.

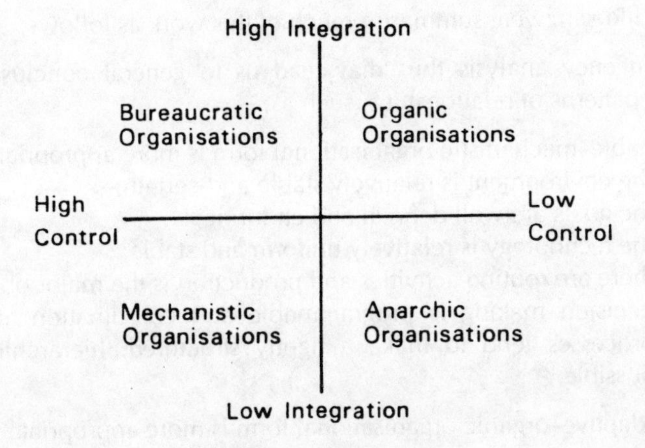

Figure 5.4 Management integration and control

Lansley *et al.* label the four types of organisation bounded by the axes as bureaucratic, organic, mechanistic and anarchic, and suggest that each type or organisation is most appropriate to different tasks and circumstances such that:

> The following groupings and related structural requirements appeared to exist in the sample:
>
> (a) General contractors. For these companies each new contract poses fresh and unfamiliar problems, creating a need for considerable flexibility with few opportunities for programming of operations. At the same time the need for coordination and teamwork is considerable. Hence the structural requirement is for high integration and low control (organic).
>
> (b) Specialist contractors. Contracts tend to follow a set pattern and operations can be largely programmed. At the same time the requirement for teamwork remains high and the appropriate form of structure is high control and high integration (bureaucratic).
>
> (c) Small works. Here also much of the work can be programmed. At the same time it is carried out by small units which can work independently. The requirement is for high control and low integration (mechanistic).
>
> (d) Companies which put most of the actual construction work out to sub-contract. These firms do not require either high control or high integration. Their activities are entrepreneurial, creative and mediated through personal relations rather than formal channels of organisation (anarchic).
>
> (Lansley *et al.*, 1977)

Kast and Rosenzweig summarise much of this work as follows:

> Contingency analysis thus may lead us to general conclusions about these patterns of relationships, such as:
>
> The stable–mechanistic organisational form is more appropriate when:
> 1. the environment is relatively stable and certain;
> 2. the goals are well defined and enduring;
> 3. the technology is relatively uniform and stable;
> 4. there are routine activities and production is the major objective;
> 5. decision making is programmable and coordination and control processes tend to make a tightly structured, hierarchical system possible.
>
> The adaptive–organic organisational form is more appropriate when:
>
> 1. the environment is relatively uncertain and turbulent;
> 2. the goals are diverse and changing;
> 3. the technology is complex and dynamic;
> 4. there are many non-routine activities in which creativity and innovation are important;
> 5. heuristic decision making processes are utilised and coordination and control occur through reciprocal adjustments. The system is less hierarchical and more flexible.
>
> (Kast and Rosenzweig, 1981, p. 116)

The environment under which urban renewal takes place is turbulent and uncertain: economic demand for land and premises is subject to significant short run variation; the level and nature of state regulation and subsidy are subject to periodic but unpredictable change. Goals are diverse and changing: each of the many groups and organisations involved will have different goals for the intervention process, some will emphasise quality in the physical environment, some the rate of return on investment, some the social impact, and these goals will change over time, for example following a change in government or market circumstances. The technology is complex and dynamic: getting anything done in urban renewal is difficult and complicated, there are many skills involved and a high level of cooordination is required. There are many non-routine activities in which creativity and innovation are important. Each situation is different: while certain activities may appear superficially repetitive (e.g. the valuation of property or the design of parking spaces) they take place in a different context each time. Each particular intervention involves differences in location; people; organisations; economic, social and technical relations.

In these circumstances there can be little doubt that the intervening organisations require adaptive–organic organisational forms with generally heuristic decision-making processes.

5.4 Market Research, Consultation and Participation in Urban Renewal

At the most basic level there is a need for organisations to know as much as possible about the market for their product or service in order that they may operate most efficiently and effectively. This knowledge can be gained through trial and error, informal consultation, formal consultation mechanisms or scientific market research. Most private producers of goods or services constantly monitor sales so as to refine products and service delivery and maximise profit potential. Not to do so would simply be poor management. Amongst the agencies concerned with public sector service delivery, including those associated with urban renewal, there seems to be a reluctance either to consult or research effectively client needs before decisions are taken or to monitor their effects afterwards.

It is an important element of the rational decision-making process that consultation and market research should take place. During the phase of objective setting and problem identification it is necessary to establish the perceptions of a variety of actors (whether individual residents, employees, service or product consumers, service or product suppliers, employers or pressure groups) as to the relative importance and severity of different problems so that appropriate priorities and solutions may be devised. During the stage of policy formulation it may again be necessary to seek specialised information and expert assistance in the analysis of situations and issues. Policy evaluation and selection has to take account of the reactions of individuals and organisations likely to be affected by the decision, and the implementation of policy may require the cooperation of other groups and organisations who are more likely to respond favourably if they have been consulted at an earlier stage in the process. The monitoring of policy effectiveness also requires consultation and market research.

Thus simply to carry out their tasks in a proper manner urban renewal agencies have to interact with their external environment. The next question is how far they should involve the external environment in decision-making processes, i.e. should they participate and why?

It has already been suggested that an adaptive–organic model of organisational form is likely to be most effective in urban renewal. This in turn implies that there will be substantial consultation within the organisation and the participation of individuals from all parts and levels of the organisation in decision making. However, it is further possible to argue that such an organisation needs to be relatively open to the external environment, having the same relaxed adaptive–organic relationship with client groups and supplying agencies as it has within its own boundaries. Such an approach might have the twin benefits of reducing uncertainty and turbulence in the external environment while the decision-making process gains from the participation of groups and organisations from all parts and levels of the urban renewal process. This provides a clear justification for moving beyond consultation towards participation, not for political reasons (although those arguments may also be compelling) but simply in the interests of more efficient and effective management.

In another argument drawn from organisation theory, Tannenbaum says:

> Hierarchy is divisive, it creates resentment, hostility and opposition. Participation reduces disaffection and increases the identification of members with the organisation...Paradoxically, through participation, management increases control by giving up some of its authority.
> (Tannenbaum, 1966, quoted in Pugh *et al.*, 1983, p. 78)

Adapting Tannenbaum's arguments it could be suggested that total control over the urban renewal process might be increased through wider participation in the decisions made by all agencies concerned. Lickert (1961) has also argued that 'participative group management' produces higher productivity, personal involvement and better industrial relations. Again it might be suggested that inter-corporate and client group participation should lead to greater individual identification with the aims and implementation of urban renewal and better relations between agencies and clients. The recent work of the Priority Estates Project for the Department of the Environment effectively illustrates the benefits of such an approach.

In a recent paper that seeks to apply lessons from good private sector management practice to the British local government context, Donnelly (1987), while acknowledging the difficulties and limitations of such transfer, develops a comparison between the eight principles of good management practice identified by Peters and Watermann (1982) and what Donnelly regards as 'the prevailing culture of local authorities'. First, Peters and Watermann's eight principles:

(i) bias towards action: innovation is encouraged and rewarded;
(ii) being close to the customer: the 'customer first' culture;
(iii) encouragement of staff autonomy and entrepreneurship;
(iv) productivity through people: management treats the workforce with care and respect;
(v) hands on, value driven: senior management has an intimate understanding of the production or service delivery process;
(vi) sticking to the knitting: concentrating on what they know and do best;
(vii) simple form, lean staff: highly decentralised organisations;
(viii) simultaneous loose–tight properties: strong central control over core values, but otherwise considerable automony of action.

By contrast British local authorities are seen by Donnelly as having the following characteristics:

(i) low corporate identity;
(ii) lack of trust between senior officers and the rank and file;
(iii) lack of shared values between officers and councillors;
(iv) innovation at lower level of the organisation is not expected;
(v) lack of good quality staff or training to support change;
(vi) access to information is not shared throughout the organisation;
(vii) access to real resources is hard to come by;
(viii) management styles often aloof, authoritarian and status conscious;
(ix) high level of alienation of junior staff;
(x) feedback from communities served neither actively sought nor given a high profile in planning the authorities' activities.

While these two lists are not directly comparable it is clear that in Donnelly's view local government practices, including their activities in urban renewal, are almost the antithesis of good management. Naturally there are exceptions, some very notable, but as a general comment, particularly in relation to management style and community feedback, much of what Donnelly says has the ring of truth.

Some lessons in modern approaches to community participation and the management of urban renewal can be learned from recent experience in the Netherlands.

5.5 Community Participation: The Dutch Experience

Amsterdam

Since the early 1970s there has been a remarkable development of community involvement in Dutch urban renewal. One of the earliest and most influential experiences was in the Dapperbuurt area of Amsterdam where the local residents rejected the city council's comprehensive redevelopment proposals arguing that they were happy in the area and did not want the overspill, housing form or higher rents that were the inevitable accompaniments to comprehensive renewal. An action committee of residents, shopkeepers and local business people was formed to argue for retention of existing streets and building lines, low rent property and citizen participation. At first there was little response from the council and a long struggle ensued in which the residents on the one hand harrassed the 'official' process and on the other hand prepared positive alternative proposals. By 1976 the plan for the area was changed. From that time the city council has come to accept a process of gradual renewal through street level citizen participation.

> In this new process of gradual renewal clearance is one of the tools of urban renewal – not alternative to it. The difference is that the residents of the blocks are:
>
> - involved in the decision to clear
> - involved in saying where they want to live instead
> - (able to) choose the design and layout of their new homes
> - and the old block is not cleared until the new homes are ready.
>
> (Community Forum, 1987 , p. 7)

The residents of Dapperbuurt organised themselves originally; now the city provides them with a superb office. Residents' groups around the city have full-time paid workers and other resources to enable them to do their job properly. The finance for these groups is committed long term by the city and there is no holding to ransom by threat of grant removal.

> The reason, it seems, is that those in power believe that it is better to give the resources to make it possible for right decisions to be made in the first place, than to withhold the resources and pay later with the trouble that results from making bad decisions.
>
> (Community Forum, 1987, p.8)

It also seems to be the case that the giving of resources to residents reduces antagonism and makes for a more productive dialogue between the parties. Another influential factor is that under Dutch law there is a right of appeal on planning decisions without a time limit on the decision. This can delay proposals indefinitely so it seems to pay to consult first. Figure 5.5 shows the resulting pattern of renewal in Dapperbuurt.

Rotterdam

In Rotterdam in recent years developments in the formal recognition and integration of citizen participation as a normal element in the urban renewal process have reached a very high level of sophistication. Under a 1974 policy decision the City Council sought to establish a vigorous urban renewal policy to deal with housing and environmental problems in the worst parts of the city. The Council identified eleven urban renewal areas comprising about 60,000 dwellings (25 per cent of the city) in which its objectives were:

(i) to improve the quality of buildings and the environment;
(ii) to retain the character of neighbourhoods;
(iii) to continue to provide housing within affordable rent limits.

Having established the broad parameters of policy the Council chose to decentralise much of the subsequent decision-making regarding the imple-

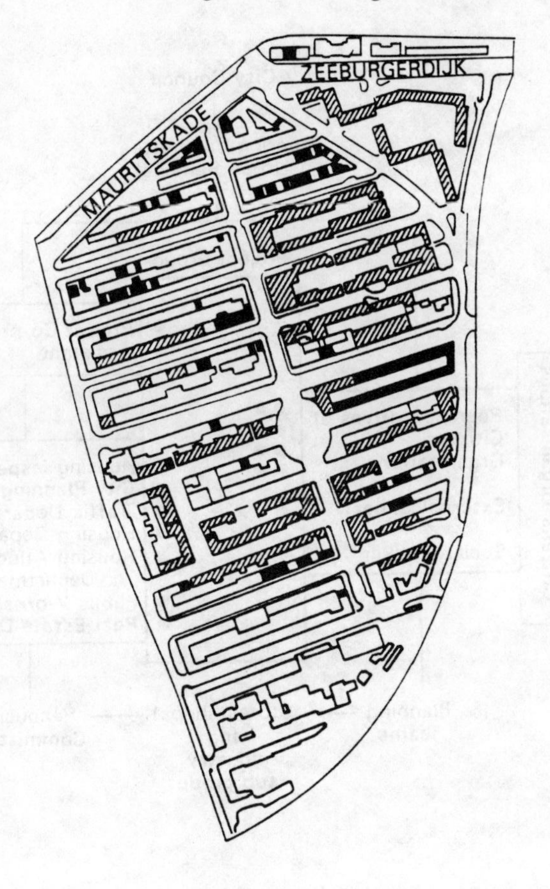

Refurbished

Redeveloped post 1970

Redeveloped before 1970

Source: after Sociale Woningbouw 68-86 Amsterdam City Council

Figure 5.5 Dapperbuurt, Amsterdam, showing the piecemeal nature of redevelopment

mentation of renewal to project teams in the areas. Figure 5.6 shows the composition of these area teams and their relationship with the central City Council administration.

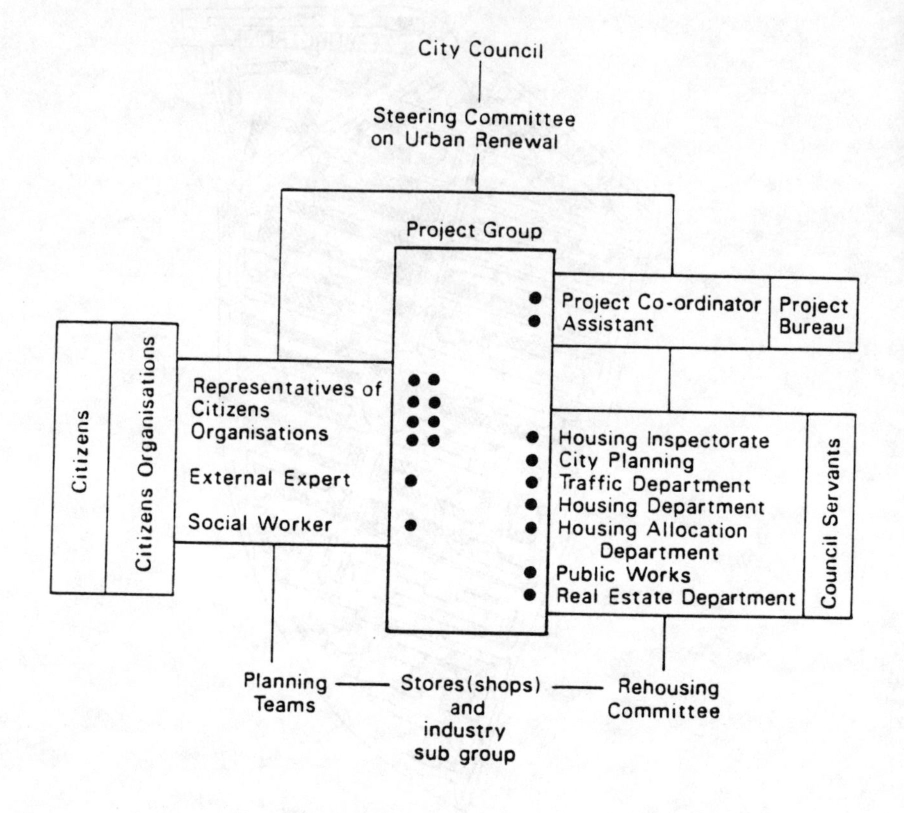

Figure 5.6 The management of urban renewal in Rotterdam

Each dot in Figure 5.6 represents one place on the decision-making team: ten for local people and their two expert advisors (a community development worker and an urban renewal expert, both paid by the City Council but working exclusively for the local community); and nine places for the City Council officials. The presence of the locally based experts enables City Council proposals to be expertly tested by the community while at the same time allowing them to develop professionally worked out and presented ideas that can be seriously considered by the City. Each city officer working on a local project team is mandated to make decisions without constant refence back to the centre. Meetings between local residents and the city are usually held within the neighbourhood rather than at City Hall. This is felt to have important psychological as well as practical benefits for the participation process. Supporting policies used by the city in the renewal process include a substantial municipalisation programme through which the Council purchases many improvable properties and the allocation of substantial and stable municipal funding to housing repair, improvement and replacement programmes.

The central city administration monitors and keeps control of the process by means of a fortnightly Urban Renewal Committee to which each group reports in turn (on a cycle!). In this way the Council is kept informed of progress and problems, gets direct feedback from the community and is able to make better informed decisions itself. It has the benefit of giving citizens direct access to the main City Council decision-making process and is also said to strengthen the hand of the city in its dealings with central government as councillors are able to speak with more authority knowing that they can truly reflect the views of the local population.

By 1985 some 36,000 dwellings in these areas had been treated and many local environmental improvements carried out. However, these statements hide the true success and stature of Rotterdam's achievements, for the figures say nothing about the very high quality of renewal and environmental design that has been achieved. Nor do they describe Rotterdam's bigger lesson and contribution to planning theory in showing that a major and sustained process of citizen participation in urban renewal can and does work to the benefit of all parties and is not something for city councils to fear but something to embrace as a way of producing better urban renewal.

5.6 Conclusions

While much if not all decision making in urban renewal can be identified as rational, it is seldom if ever part of any comprehensive process of moving towards societially agreed goals and objectives; rather there are a multiplicity of decisions taken by large numbers of agencies and organisations which may be individually rational in relation to private objectives but in conflict or counterproductive in relation to broader considerations. It is into this situation that the state or community agencies of urban renewal must intervene and recognise that they have often to work at the level of the lowest common denominator and proceed incrementally through individually negotiated steps.

This is not to say that there can be no direction or strategy. Clearly it is possible to establish long term strategies and to make localised day-to-day tactical decisions within that framework. But these strategies must recognise the nature of conflicting interests acting upon cities and develop accordingly.

The decision-making environment in urban renewal is turbulent and uncertain and agencies need to acknowledge this in their structure and operations. This is likely to mean that adaptive–organic organisation with heuristic decision-making processes will be most appropriate.

Above all there seem to be strong practical arguments for the opening up of urban renewal decision making to consultation and participation processes. Both the interpretation of management theory and the practical experience of many cities, notably in the Netherlands, suggests that participation enhances the community benefit from urban renewal.

6 Urban Design for Urban Renewal

Urban renewal involves changes in the physical fabric of cities. These changes can make the city function better or worse and they can make the city look better or worse. Each change can be initiated without regard or with inadequate response to these two issues or it can be initiated in such a way as to improve the functioning and the look of the city. In this context the word functioning refers to the way the city works as a physical unit: the efficiency of its physical infrastructure; and by the look of the city we mean the extent to which the city is aesthetically legible and pleasing to the eye. Urban design is a process of responding to these two issues when making decisions about the location and physical manifestation of investment in the built environment: it is the adaptation of building and engineering operations to functional and aesthetic ends.

In the building and rebuilding of cities a process of urban design has always been implied. Historically this process has sometimes been conscious, grandiose and comprehensive; for example in the classical city, the bastide towns, parts of Georgian London and the spa towns or Housemann's Paris; but in many more instances the process has been unconscious, implicit and piecemeal. Most towns and cities throughout history have grown in this organic fashion. Frequently the functional and aesthetic results of this organic process have justified few design criticisms until these places have been overwhelmed by sudden accelerations in the rate of urbanisation or the intrusion of non-local materials, building methods or investors without local awareness or sympathy.

Many architects, planners and other concerned authors have written about the design of urban areas, either to interpret, explain and evaluate what has gone before or to offer prescriptive methods and solutions to the problems of designing urban areas in modern circumstances and to meet modern needs. Here the contributions of a number of such writers are considered in approximately the chronological order of their work. This is not intended to be a comprehensive review of thé literature but simply an introduction to some key ideas and developments. By examining these approaches and ideas it is possible to make some tentative statements about the content and processes of urban design appropriate for urban renewal today.

6.1 The Image of the City

According to Lynch (1959) perceptions of the city as a physical entity are conditioned by and interpreted through the existence of five elements: paths, edges, districts, nodes and landmarks.

> Paths are the channels along which the observer moves ... for many people these are the predominant elements of their image. People observe the city while moving through it, and along these paths the other environmental elements are arranged.
>
> (Lynch, 1959, p.47)

A path is the route chosen by the observer. It does not necessarily confirm with the patterning and hierarchy of paths shown on a map. The observer's image of the city is determined by his or her chosen route and mode of transport: home to work by train; home to shops by car; home to school on foot. For each person the city is viewed from different routes travelled at different speeds: each observation is unique.

> Edges are the linear elements not used or considered as paths by the observer. They are the boundaries between two phases, linear breaks in continuity: shores, railroad cuts, edges of development, walls. They are lateral references rather than coordinate axes.
>
> (Lynch, 1959, p.47)

For Lynch edges are not as dominant as paths but are nevertheless important organising features helping to define generalised areas.

> Districts are medium to large sections of the city, conceived of as having two-dimensional extent, which the observer mentally enters 'inside of ', and which are recognizable from the inside, they are also used for exterior reference if visible from the outside.
>
> (Lynch, 1959, p.47)

Extreme examples would be the historic core of many European towns: *the vielle ville* in French cities or the *altstadt* in Germany; where the tightly enclosed, intense development of the aged core is self-evidently a different district from the loosely developed open extensive modern surroundings. But every town can by broken into districts by the observer. Visitors to London can identify the physically different districts of Bloomsbury, Whitehall, Kensington, Hampstead village, or the City of London.

> Nodes are points, the strategic spots in a city into which an observer can enter, and which are the intensive foci to and from which he is travelling. They may be primarily junctions, places of a break in transportation, a crossing or convergence of paths, moments of a shift from one structure to another. Or the nodes may be simply concentrations, which gain their

importance from being the condensation of some use or physical character, as a street corner hangout or an enclosed square.

(Lynch, 1959, p.47)

Landmarks are another type of point-reference but in this case the observer does not enter within them, they are external ... Their use involves singling out one element from a host of possibilities. Some landmarks are distant ones, typically seen from many angles and distances ... other landmarks primarily local, being visible only in restricted localities and from certain approaches.

(Lynch, 1959, p.48)

In London Big Ben and St Pauls are both distant and local landmarks, although as it is a relatively flat city with many high buildings these are not particularly strong landmarks when contrasted with say, the Eiffel Tower in Paris or the Metropolitan Cathedral in Liverpool. Landmarks need not simply be high buildings: the main railway station is an essential feature in orientating oneself around Amsterdam and the Puy de Dome dominates the western skyline of Clermont Ferrand. A landmark may be much smaller than this; a statue, a corner building or even a tree; it may be visible for only a couple of street blocks in any direction, nevertheless it may perform an important localised reference function.

People need paths, edges, districts, nodes and landmarks in order to understand the physical structure of the city, to find their way around, to get to know and enjoy the city. The urban renewal decision that removes a commonly used path or an important landmark is doing a disservice to the community, making it more difficult for people to relate to the city and increasing their alienation from it. Decisions that increase the legibility of the city by clarifying pathways, emphasising landmarks and maintaining edges will increase people's understanding, appreciation and enjoyment of the city as a physical entity.

6.2 British Ideas of Townscape and Conservation

Cullen (1961) refers to three concepts that he feels are important ways in which our visual reaction to the city environment is stimulated. These are through 'serial vision', the idea of 'place' and the 'content of places'.

Let us suppose that we are walking through a town: here is a straight road off which is a courtyard, at the far side of which another street leads out and bends slightly before reaching a monument ... our first view is that of the street. Upon turning into the courtyard a new view is revealed instantaneously at the point of turning, and this view remains with us while we walk across the courtyard. Leaving the courtyard we

> enter the further street. Again a new view is suddenly revealed although we are travelling at uniform speed. Finally as the road bends the monument swings into view. The significance of all this is that although the pedestrian walks through the town at a uniform speed, the scenery of towns is often revealed in a series of jerks or revelations. This we call serial vision. Examine what this means ... A long straight road has little impact because the initial view is soon digested and becomes monotonous. The human mind reacts to a contrast, to the difference between things, and when two pictures (the street and the courtyard) are in mind at the same time, a vivid contrast is felt and the town becomes visible in a deeper sense. It comes alive through the drama of juxtaposition.
>
> (Cullen, 1961, p. 9)

The further implication of this idea of serial vision is that the city now falls into two elements: the present view, and the emerging view; and each has to be taken into account in this dynamic concept of urban design. It will be noted that there are distinct parallels with Lynch's idea of pathways and the way the image of the city changes as we move along them.

Cullen's second concept is that of place and is concerned with:

> our reactions to the position of our body in its environment. It means, for instance, that when you go into a room you utter to yourself the unspoken words 'I am outside it, I am entering it, I am in the middle of it'.
>
> (Cullen, 1961, p. 9)

This same idea can apply to the street or square: a definite feeling of being outside, entering and leaving. Once the town can be divided into identifiable places that one is either in, not in, entering or leaving, it becomes possible to talk of 'here' and 'there' and, as Cullen points out: some of the greatest townscape effects are created by a skilful relationship between the two.

In the content of places Cullen is concerned with the fabric of towns and cities: the colour, texture, scale, style, character, personality and uniqueness. Most towns are a jumble of building types, styles and ages. While our first reaction might be to say that the whole would look better if replaced by new structures with symmetry, balance and standardisation, Cullen suggests that such a conformist city would be boring and that it is the very variety of the historic–organic city that gives it visual charm and attraction. Yet this organic growth has not been as diverse as we might think (at least before these places became overwhelmed by sudden accelerations in the rate of urbanisation or the intrusion of non-local materials, building methods or investors without local awareness or sympathy). There was an unconscious, unspoken 'agreement to differ within a recognised tolerance of behaviour', i.e. this variety evolved within an overall unifying set of rules or criteria which provided the framework for organic growth. That framework sets the conventions in the colour, texture, scale, style and so forth that dominate building in that particular town. Where conventions of colour or materials might be broken the building would probably still conform in proportion, scale and style and remain recognisable as part of that town (and no other). So the fabric of the town is not uniform but an interplay of 'this and that'.

> Thus we have motion, position and content: or to put it another way: serial vision, here and there, this and that ... all that remains is to join them together into a new pattern created by the warmth and power and vitality of human imagination so that we build the home of man. This is the theory of the game, the background. In fact the most difficult part lies ahead, the Art of Playing.
>
> (Cullen, 1961, p. 12)

Writing in 1968 at the culmination of over thirty years scholarship in urban design, Sharp summed up his approach in a book called *Town and Townscape*.

> Never before have there been so many different influences for change operating so strongly upon (towns) at the same time. And never before has the tempo of possible change been so swift ... The influences for change are partly social needs, partly mere fashion.
>
> (Sharp, 1968, p. 1)

Sharp highlights the motor vehicle and changing land use demands as major social forces but also makes the point that:

> It is through the architects and their clients that the influence of mere fashion has had, and is having, its effect – fashion in street buildings that either lack street perspective or are hopelessly at odds with well established building rhythms; in tower buildings that are wholly out of scale and character with the towns over which they exert architectural tyranny. In these new buildings all previous acceptance of something like a collective discipline has been rejected. It has been rejected through an architectural arrogance in which the general character of the town or street is considered of no importance compared with the intoxication of self-assertion and self-advertisement.
>
> (Sharp, 1968, p. 3)

While Sharp was writing towards the end of an era in which massive slum clearance, town centre redevelopment and the influence of the 'modern movement' in architecture were dominant, and before the conservation movement had become established, his argument retains validity today. He emphasises a concern that new building within existing towns and cities should accept and be designed within the discipline imposed by the existing urban fabric both in terms of the scale and chararter of existing buildings and the rhythm and perspective of existing streets and spaces. He makes a similar point to Cullen that what makes historic, organically evolved towns aesthetically interesting is often variety in building form, scale and materials but:

> variety within the same kind, variety within an established rhythm, variety ... within a broad unity of character.
>
> (Sharp, 1968, p. 13)

In the following year Worskett, a former Ministry of Housing and Local Government advisor, wrote a book called *The Character of Towns* in which he sought to show how conservation policy ought to take account of townscape qualities in addition to mere building preservation. Worskett saw conservation policy evolving out of two parallel sets of concerns. Beginning with a visual and historical survey of the area concerned, work should proceed, on the one hand, to identify historic buildings and archeological features and to establish aims or principles governing their preservation; while on the other hand there needs to be an assessment of townscape qualities and a visual discipline established for the design of changes to the physical environment. It is through the amalgamation of these two elements that conservation policy is formed.

Worskett sub-divides the development of a visual discipline for conservation design into four parts:

(i) town–landscape relationship: the appearance of the town as seen from the countryside and the importance of defining and maintaining urban boundaries;
(ii) high buildings policy: both the maintenance of skylines and focal points and the exploitation of opportunities to enhance skylines by adding new focal points;
(iii) townscape discipline: both the maintenance of those qualities of space and layout that create a local discipline and the exploitation of opportunities to enhance a scene through new building, renovation, landscaping, etc.;
(iv) design of infilling: the architectural effect of new buildings inserted into existing streets, in which he sees continuity as being dependent upon a number of qualities including: building line; building height and skyline; width of the building unit; quality of detailing and use of materials; and the proportions of window to wall area.

Worskett also points to the tremendous visual benefits that can be achieved through simple 'tidying up' of areas by the removal of unnecessary street signs, replacement of unsuitable street furniture, renovation of pavements and floorscape and improvements to existing building facades.

Turning more towards the practicalities of policy formulation and implementation Workskett identifies a number of issues that must be addressed.

(i) The selection of areas and the definition of boundaries.

> The greatest danger lies in the isolation of such areas from the organic form and structure of the town. To continue to have meaning a conservation area must function and appear to function as an integral part of the town as a whole.
>
> (Worskett, 1969, p.46)

(ii) Establishing priorities: is it better to concentrate effort on conserving a few exceptional areas or to seek a more general widespread implementation of conservation principles?

(iii) The acknowledgement that if conservation implies restriction on growth in one area then provision has to be made for growth elsewhere. One might contrast the City of London (where, until recently in Docklands, no alternative growth points were available and effective conservation over a wide area was an impossibility), with say, Oxford (where central area conservation has been coupled with the accommodation of some growth around the peripheral outer ring road). One might also look at French cities such as Paris, where growth at La Défense and Marne la Vallée has permitted tough conservation rules to be applied within the historic city, or Lyons, where by syphoning off development pressure into Part-Dieu the city has very successfully conserved the nineteenth-century commercial and residential centre.

(iv) The understanding that conservation and building preservation policies can only operate within the realities of building economics. The occupation of buildings by appropriate uses must remain profitable under the limitations and obligations of preservation policy or be subsidised or put to some social or public sector use.

(v) Worskett also stressed the importance of public relations and consultation in the conservation process.

6.3 Design Guidance and Development Control

In an effort to improve the quality of building design and townscape awareness both the Department of the Environment and a number of local authorities began to issue 'good practice' design guidance. One of the earliest and most well known of these was the Essex Design Guide (Essex CC, 1973) with a foreword written by the Secretary of State.

Although not specifically concerned with urban renewal and limited to the design of residential areas only, this is nevertheless an important landmark in raising the sophistication of local planning control over detailed site planning and design. As the Guide states:

> What constitutes good design has far too often been written off as a matter of opinion or taste with a subsequent cheapening of the visual environment. By setting out a clearly related structure of planning and design policies, it is hoped that better housing schemes and a greater consistency in the granting of planning permissions will result.
>
> (Essex CC, 1973, p. 19)

A number of these policies are quoted below since many of them have subsequently been adopted or repeated in a modified form by other authorities and can be regarded as typical of modern good practice in housing design and layout, whether in greenfield or urban renewal situations.

Physical criteria

> New housing should normally be designed to at least the minimum 'Parker Morris' space standards
>
> (p. 26)

> a feature that it is sadly increasingly difficult to achieve in new housing today.

> Where homes are proposed smaller than the recommended space standards ... then a satisfactory means of extending the property should be indicated
>
> (p. 28)

> On the 'public zone' side of the dwelling, a reasonable degree of internal privacy should be achieved ... and on the 'private zone' side a high degree of internal privacy shall be provided ...
>
> (pp. 30–31)

> Houses shall have a minimum 'private zone' garden area of $100m^2$ (different policies apply to flats and old persons' bungalows)
>
> (p. 35)

> Within new housing areas pedestrian movement shall be made convenient, safe and pleasant, by the provision of carefully positioned and well designed 'pedestrian spine routes' and 'local access footpaths'
>
> (p. 38)

> Within new housing areas, vehicular movement shall be made convenient, safe and pleasant by the provision of:
>
> (i) a road system segregated from through traffic;
> (ii) a clearly defined hierarchy of routes, with roads differing in size, alignment and standard according to the designed speed and volume of traffic they are intended to carry
>
> (p. 40)

Visual criteria

The Guide notes the distinction between the truly rural situation in which the key visual feature is 'landscape containing buildings' and the truly urban situation in which the key visual feature is 'buildings containing space', and condemns much modern 'suburban' development as falling between these two stools and consequently failing to provide a visually satisfactory environment. In might be noted in passing that most of the recent (late 1980s) wave of private housebuilding and some public sector building in the innermost parts of British cities is falling into this very same visual trap. The Guide suggests that:

> the prime underlying principle of all urban places is the enclosure of space by buildings. If space is not satisfactorily enclosed, an attractive urban place cannot be achieved.
>
> (Essex CC, 1973, p. 64)

(But see the comments by Hillier below.)

To this end the Guide advises that:

> new residential areas should consist of a series of satisfactorily enclosed, contrasting spaces, each related to the human scale
>
> (p. 65)

It distinguishes between dynamic spaces (e.g. streets) in which satisfactory enclosure can be achieved by effective height–width ratios of between 1:1 and 1:2.5, and static spaces (e.g. squares) where a ratio of around 1:4 would, in their view, provide a harmonious relationship between the elements.

Trees are identified as performing two potentially important functions within urban spaces: visually to complete built frontages; and to modify built spaces (e.g. to break up a large space into smaller, visually more satisfactory, units).

With regard to the architectural quality of buildings it is suggested that:

> Individual buildings shall be well designed in themselves and have regard for their setting by;
>
> (i) being designed to form part of the larger composition of the area in which it is situated;
> (ii) using suitable external materials for the location in which it is situated;
> (iii) the volumes making up the block form of the building being proportioned and related to form a satisfactory composition;
> (iv) the external materials being used in a visually appropriate manner;
> (v) the fenestration being well proportioned and well related within the elevation and also being sympathetic to adjacent buildings;

(vi) architectural detailing being used to reinforce the character required by the design and its location.

(p. 74)

Following the introduction and widespread adoption of General Improvement Area policies by local authorities, the Department of the Environment began publishing a series of 'Area Improvement Notes' during the early 1970s.

These advice notes covered a variety of topics including housing condition surveys within potential General Improvement Areas, environmental improvements, housing improvement and conversion, public participation in General Improvement Areas, car parking and other topics. In spite of their age many of these notes contain advice still of relevance to area improvement. One such note dealt with the topic of 'Environmental Design' (DOE, 1972). That note described the environmental problems faced in a number of 'typical' improvement areas and discussed the solutions adopted by local authorities, paying particular attention to traffic management, garaging and parking, pedestrian areas, children's play, appearance and new infill buildings. The discussion focussed upon the design aspects of environmental improvements and the difficulty in striking a balance between conflicting objectives. While discussing the pros and cons of various solutions the note concluded with no 'rules of thumb' but a recognition that the problems of individual GIAs varied considerably and therefore required bespoke solutions. In other words a 'contingency approach' was required. Nevertheless common issues needing resolution were: the diversion of through traffic out of the area; the accommodation of essential non-residential traffic; residents and visitors' access and parking; possible 'pedestrianisation' (to improve the amenity of dwellings, or to improve the street scene with planting or paving, or to redirect traffic, or even to increase the site area for redevelopment); the provision of pathways and safe pedestrian routes to open spaces, playgrounds, bus stops, schools and shops, etc. It was also noted that many GIAs comprised monotonous byelaw streets and ways were suggested by which this monotony might be broken up and relieved, e.g. through judicious infill building, planting or treatment of facades.

Figure 6.1 illustrates one of the examples studied by the Department of the Environment: Camdentown in Gosport. This was a General Improvement Area comprising 633 dwellings, 55 shops and some light industry. The plan shows the proposals for a new section of road, a playground, a continuous central pedestrian path, grouped garages, some limited demolition and infilling and new planting and landscaping.

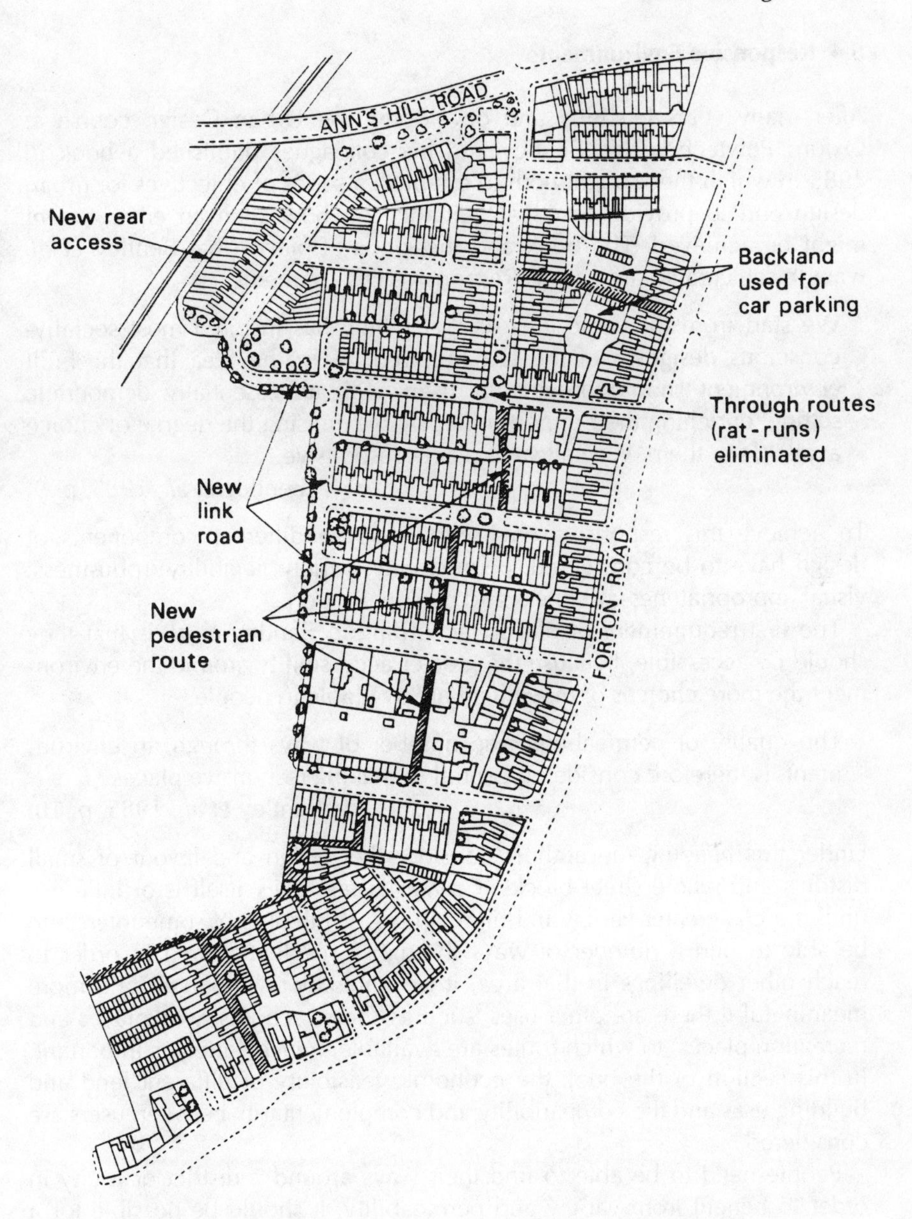

Source: Adapted from a diagram in DOE Area Improvement Note № 5 HMSO 1972

Figure 6.1 Camdentown General Improvement Area, Gosport

6.4 Responsive Environments

After many years teaching and developing the Urban Design course at Oxford Polytechnic Ian Bentley and his colleagues published a book in 1985 in which they sought both to establish a series of objectives for urban design and to provide practical guidance on how such an environment might be achieved. The book and the kinds of places that Bentley *et al.* want to see created are called 'Responsive Environments'.

> We start from the same idea as that which has inspired most socially-conscious designers of the last hundred years: the idea that the built environment should provide its users with an essentially democratic setting, enriching their opportunities by maximising the degree of choice available to them. We call such places responsive.
>
> (Bentley *et al.*, 1985, p. 9)

To achieve this responsive environment seven different components of design have to be considered: permeability; variety; legibility; robustness; visual appropriateness; richness and personalisation.

The first requirement of places, according to Bentley *et al.* is that they should be accessible, because the greater accessibility around the environment the more choices of movement are available to people.

> The quality of permeability, the number of ways through an environment, is therefore considered central to making responsive places.
>
> (Bentley *et al.*, 1985, p. 10)

Under this heading the authors consider the design and layout of small districts and whole street blocks. Greater accessibility itself is of little use until there is greater variety in building uses. While it is of some interest to be able to find a number of ways through a residential area in order to reach other dwellings in that area, it becomes more important and more meaningful if there are other uses (such as shops, schools, workplaces and recreation places) to which routes are available; hence variety is important. In this section of the book the economic feasibility of different land and building uses and the compatibility and complementarity between users are considered.

People need to be able to find their ways around a district or a city in order to benefit from variety and permeability. It should be possible for a person to read the city in the sense of being able to identify easily what type of area they are in, to recognise features and routes within that area. The city should be legible.

> The tentative network of links and uses already established now takes on three-dimensional form, as the elements which give perceptual structure to the place are brought into the process of design. As part of this

process, routes and their junctions are differentiated from one another by designing them with differing qualities of spatial enclosure. By this stage, therefore, the designer is involved in making tentative decisions about the volumes of the buildings which enclose the public spaces.

(Bentley *et al.*, 1985, p. 10)

In their subsequent analysis of urban legibility the authors draw heavily upon the work and approach of Lynch (discussed earlier) in referring to the importance of being able to identify paths, nodes, landmarks, edges and districts.

In a usage of terminology that differs from our early references to flexibility and robustness in the discussion of the planning process in Chapter 4, it is suggested that a desirable state to be achieved is one in which individual buildings or places may be put to more than one use. This flexibility is curiously referred to as 'robustness'. By way of example a contrast may be drawn between the flexibility of the late-Victorian bylaw terraced house and a flat in a multi-storey block. The terraced house might be used as a single dwelling; a pair of flats; partly or wholly converted into a shop, office or workshop; and any single unit may be demolished and replaced, extended, refurbished or converted without damage to the rest of the terrace. The flat has little flexibility in use and cannot be demolished or converted without damage to the rest of the block. The heading robustness therefore includes detailed discussion of the design of buildings: building depth, access, height, configuration, internal organisation, nature and use of outdoor spaces and so forth.

People interpret the meaning of environments, like the degree of legibility, whether designers want them to or not; therefore:

a place has visual appropriateness when these meanings help to make people aware of the choices offered by the qualities we have already discussed.

(Bentley *et al.*, 1985, p. 10)

At this level the interpretation of a building or area's function and importance will be triggered by a series of visual cues, or clues, given by the physical appearance of the place: vertical and horizontal rhythms; skylines, building height and scale; wall detailing (materials, colour and texture); windows, doors and ground level details.

The decisions so far taken in this hierarchical approach to urban design:

still leave room for manoeuvre at the most detailed level of design. We must make the remaining decisions in ways which increase the choice of sense–experiences which users can enjoy. This further level of choice is called richness ... by this stage, we are dealing with the smallest details of the project. We must decide whereabouts in the scheme to provide

> richness, both visual and non-visual, and select appropriate materials and constructional techniques.
>
> (Bentley *et al.*, 1985, p. 11)

The final stage in the process, known as personalisation, is the point at which design moves away from the professional designer of the 'public realm' to decisions made by individual owners and users about the decoration and 'garnishing' of buildings and spaces. The authors stress the desirability of designing to support personalisation while trying to ensure that its results do not detract from any public role the building or space may have.

6.5 More Recent Ideas and Approaches

For many years Christopher Alexander and his associates have been striving towards a better understanding of the design process. In 1987 they published a short book that represented the distillation of a series of discussions and a lengthy experiment which suggested the formulation of an entirely new way of looking at urban design; indeed, so much so that the team felt justified in entitling their findings as 'a *new* theory of urban design'.

What they suggest is that the primary aim of urban design is to create a city which is *whole*, i.e. which is always seen as a complete entity at any stage of its growth and which can be seen to be a complete whole at any time. That is to say that like a growing child or a growing plant the city should always be a complete entity for its appropriate stage of growth.

> We believe that the task of creating wholeness in the city can only be dealt with as a PROCESS. It cannot be solved by design alone, but only when the process by which the city gets its form is fundamentally changed.
>
> (Alexander *et al.*, 1987, p.30)

Thus Alexander and his team are as concerned to propose a new process of urban design as much as they are concerned to propose what the content should be. They set themselves the task of learning the laws, or rules, which produce wholeness in a city. From this they were faced with the question: what kinds of laws and at how many levels are needed to create a growing whole city or part of a city? In conclusion they formulated one single overriding rule which required that: 'Every increment of construction must be made in such a way as to heal the city'. This word 'heal' is used in the sense 'to make whole'. In order to translate this aim into a working practical theory a series of secondary rules are proposed. The major ones are outlined below:

1. Piecemeal growth
In addition to advocating piecemeal growth the team further suggest that no single building increment should be too large; that there should be a reasonable mixture of project sizes at all times; and, that there should be a reasonable distribution of functions (building uses) at each stage of growth.

2. The growth of larger wholes
Every building project should help to form at least one larger whole within the city. This means that each project should consciously add to a street, help create a place or square, as a kind of adaptive–organic response to the desire to improve urban structure. They suggest that it is unlikely that any one building project will itself create a new public space but an awareness of the possibilities to contribute to the creation of exterior spaces recongnises that each new development can make a positive contribution to this process.

3. Visions
This rule defines the content and character of the individual increments. If each building project is going to contribute to 'wholeness' then it is important that, in terms of physical and aesthetic qualities, it must grow, naturally and directly, from what is already there. As Cullen (1961) has said, there is in each town a unifying set of rules or criteria provided by the existing buildings that create a framework for organic growth.

Indeed, although Alexander's approach is much more theoretically rigorous than the rather subjective approaches of Cullen, Sharp or Worskett it is difficult to see that it really justifies the term 'a new theory'. However it is a useful support for the growing argument in favour of piecemeal, small scale, aesthetically sympathetic steps in urban renewal.

For a number of years Bill Hillier from University College, London has been trying to develop the analysis of townscape by moving it away from a subjective appreciation of forms and spaces towards a more scientific analysis of existing towns to find out why and in what ways some spatial forms 'work' better than others.

Whereas many modern writers and designers have argued in favour of 'enclosure' as one of the most important concepts in townscape, Hillier suggests that the concept has been misunderstood and misapplied in many modern 'planned' developments. By analysing the spatial structure of a number of historic towns with a traditional urban pattern he came to the view (here referring specifically to Apt in France but the point is a general one) that:

> First, very little of the open space can be described as enclosed in the localized sense ... Admittedly, all parts of the open space are shaped and

> defined in relation to building entrances; but, equally, all parts are related by lines of sight and access to the larger scale space structure. Even the 'squares' which are the most obviously 'enclosed' spaces ... have the important additional property of being also strategic spaces from which a good deal of the larger scale space structure of the town can be seen.
>
> (Hillier, 1988, p. 65)

He goes on to consider why it is easier to find one's way around the traditional town, with its apparent lack of order and symmetry, why it is more intelligible. He suggests that important keys to this understanding are the idea of the deformed grid. This grid is the pattern of spaces formed by islands of outward facing buildings. In the traditional town Hillier found two common features. The first was that many of the spaces created are 'convex spaces', which means that all points and all building entrances can be seen, identified and be subject to surveillance from any other point or entrance within that space. The second point is that these convex spaces are also related to each other by visibility and access so that a person standing in any one space is aware of a whole series of connecting spaces, and by moving through to the furthest visible space a whole series of further spaces is revealed. Thus each space can be understood in itself and in its relationship with other parts of the town.

From this Hillier shows how it is possible to construct a map by drawing the fewest and longest straight lines which pass through all the convex spaces in the town. The fewer lines that we have to pass along in order to get from any one point in the town to any other point the higher the level of 'integration'. Here integration really means the degree of visibility and accessibility between places. Further:

> the fewer intervening lines which need to be passed through to go from a line to every other line, then the more integrating that line; the more intervening lines, the more segregating.
>
> (Hillier, 1988, p. 69)

By identifying the most integrating lines it is possible to determine which 'streets' are the most important to the structure of the town. Hillier has found that in most traditional towns this structure takes the form of a 'deformed wheel' with important streets or spaces at the centre, forming key radial routes, and around parts of the rim. This hierarchy of intergration separates more private segregated areas from more public integrated areas. Thus strangers can find their way easily through the town, while at the same time they are following routes which many other people will follow and so their chances of encounter and surveillance are higher. Hillier goes on to correlate higher encounter rates with lower crime rates, as Jacobs (1961) had intuitively done nearly thirty years earlier.

Thus Hillier concludes that in the design of new or renewed neighbourhoods two features are particularly important to achieve: convex spaces, so that surveillance is maximised; and, highly integrated layouts, so that encounter rates are maximised. To facilitate this process a series of rules of thumb are set out in Hillier (1988, pp. 86–88).

6.6 Exploiting the Physical Potential of Buildings and Areas

In a recent article MacDonald (1989a) has examined the potential for using the existing built fabric of a large area of derelict docklands and their industrial hinterland in Liverpool as the basis for achieving the simultaneous improvement in urban design and industrial regeneration. MacDonald's approach is based upon the careful analysis of the potential uses for existing vacant premises; recognising and taking account of local ideas and skills; and working within the general tide of economic regeneration coming from the mainstream of public and private sector development agencies. He comments that:

> The present climate of building rehabilitation is now an accepted approach to dealing with many problems of inner city redevelopment. It is no longer a fringe activity to be concerned with the saving and re-use of old buildings, it is the economic reality in the deprived inner cities, where capital investment in buildings is severely limited ... However it is perhaps in the field of industrial regeneration that Liverpool has fallen behind as housing and environmental regeneration tends to have taken priority over long term industrial regeneration. There is a clear need, not the least in the North Docklands, for improvements in attitudes towards the local industries and industrial environment ... It would appear that there is a need for an inventory of all remaining buildings in the North Docklands neighbourhood. An assessment needs to be made of the buildings suitable for low cost conversion. Such an investigation into the existing building types and remaining uses, in terms of space needs and feasible rent levels, could be used to encourage existing industry to remain and grow, whilst at the same time allowing changes to occur to achieve other objectives in the field of community enterprise.
>
> (MacDonald, 1989a, p. 23)

To stimulate ideas MacDonald produced a diagrammatic representation of existing structures, proposals and potentials as shown in Figure 6.2.

Since the mid-seventies there have been many schemes sponsored by a variety of agencies seeking to exploit the regeneration potential of the existing urban fabric in many towns and cities across the country. These range from industrial improvement areas (see Chapter 7), former docklands, city centres and shopping areas to run-down residential areas and the

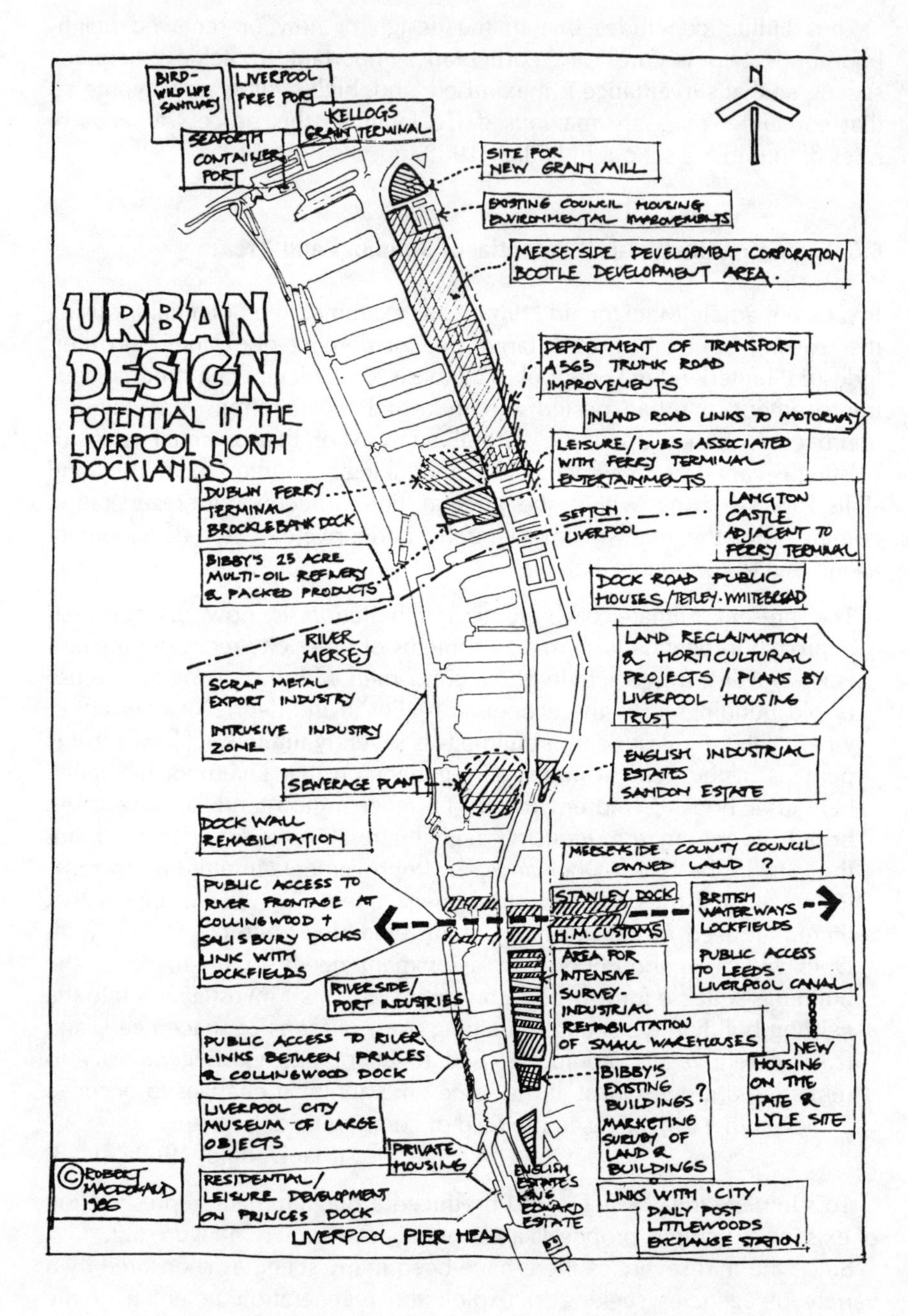

Figure 6.2 Physical potential in the North Docks, Liverpool

urban fringe. In 1988 the Department of the Environment published an analysis of this kind of urban regeneration activity under the title: *Improving Urban Areas* (JURUE, 1988). Their conclusion for good practice can be summarised under four main headings.

1. Getting started

It is argued that the initiation of projects requires imagination and commitment and that they work best when responsive to and working with grassroots pressure for improvements, when objectives are clearly set and prioritised, and when they complement existing initiatives.

2. Maintaining momentum

The employment of a full-time project officer (a project manager), the direct involvement of relevant influential decision makers, delegation of responsibility to key officers, the involvement of the local community, and the coordination of key agencies are all regarded as essential to project success. It is suggested that the future needs for project maintenance should be anticipated and funding arrangements secured at an early stage. Establishing confidence is seen as a critical ingredient of success and can be developed through demonstration projects, enthusiastic promotion and appropriate use of the media. Confidence is important as the participation of the private sector is seen as vital to the success of most schemes.

3. Critical choices

The inclusion of a variety of easy (cheap) and more complex (expensive) projects, reflecting local priorities, is seen as important. It is further suggested that project areas should have a clear distinctive identity or character; that projects should be realistic in scale and timetabling.

4. Financial support

It is argued that financial support should be established from a wide variety of sources so as to increase the robustness of the project, and that the early commitment of public sector resources should be used to stimulate and encourage private sector investment and participation.

In summary good practice is thought to depend upon the early involvement of local community interests, the establishment of clear objectives and priorities, good management and coordination between agencies, exploitation of the special local features and character of areas, and the creative combination of resources from a variety of sources.

6.7 The Healthy City

Since 1986 the World Health Organisation has been the sponsor of a multi-pronged action–research–promotional project to lift the quality of health in Western cities. This is no ordinary health promotion campaign but a major attempt to consider how the present city works against good health and how the economic structure and physical design of the city might be changed in the interests of better health for all. Work is progressing along a number of fronts simultaneously: action projects within individual participating cities, research projects, local and international conferences, and media publicity.

The justification for a review of relationships between cities and health is derived from the writings of various planning theorists, architects, medical writers and pressure groups. The common strands that emerge from this work are: the emphasis on the need to relate the development of the city to its natural surroundings both in terms of ecological balance and landscaping relationships; the idea of energy saving and pollution reduction; increasing personal safety through environmental design; increasing the vitality and mix of uses within districts; and working through the participation of local people. (MacDonald, 1989b, p. 4)

From this general approach a number of themes are proposed by the healthy cities campaign for the guidance and encouragement of urban designers:

1. Urban design projects that are economically based in the local and decentralised city and which enable people to organise locally;
2. Urban design projects which are less wasteful, less ecologically damaging and more conserving of resources;
3. Urban design projects which emphasise recycling of local resources and skills whilst reducing dependence on imports of energy and products;
4. Urban design projects which generate new forms of urban industry, production, city farms and local heat and power systems;
5. Urban design which cuts across professional and sectoral boundaries of employment, commercial, academic, cultural and leisure activity, art and science;
6. Urban design projects which develop the role of the informational technologies in the support of community initiatives;
7. Urban design projects which expand the role of community organisations, as the basis of other initiatives, into the wider debate towards the healthy European City.

(MacDonald, 1989b, p. 5)

6.8 Conclusions

Once the design professions had thrown off the crushing inhumanity of the Corbusier- and modern-movement-inspired approach to urban design that gave us high-rise housing, town centre 'reconstruction' and an arrogance that almost completely disregarded history, continuity and local communities, there emerged from the sixties onwards a fairly consistent theory of urban design practised across Britain, Western Europe and to some extent in North America.

The essence of this approach is contained in a recognition of the indispensability of human scale in urban design; the need for continuity with the past which can be achieved through the retention of landmarks, gradual renewal and a respect for traditional local building form, styles and materials; the importance of enclosure, although subject to different interpretations; the degree of personal safety brought to an area through a mixture of uses and planning for the presence of and surveillance by people.

More recently there has been a growing appreciation of the necessity for designers to understand the economic limitations and to exploit the economic potential of areas; to appreciate the complex managerial tasks involved in the implementation of urban design and to recognise the strength of arguments in favour of local community participation and the consequent de-mystification of the design process.

Most recently it seems that a new, more ecologically aware approach to urban design is emerging. This approach requires new ways of thinking, the breaking down of professional and disciplinary barriers, the understanding of complex relationships between the natural environment, economics and the man made environment, and it implies a more cooperative and less selfish or egocentric approach to design and development than hitherto.

And yet in the practice of urban design and renewal, in terms of what is being achieved on the ground, we still seem to be twenty years behind these advances in theory, if indeed these theories carry any weight at all. Why should this be the case?

Firstly, a commonly expressed argument is that there is a lack of design appreciation in this country and that this has somehow led to the acceptance of low standards of urban design. But this does not really stand up. There is no reason to suppose that our forefathers in Georgian times or the yeomen who inhabited the now much admired villages of Suffolk or the Cotswolds had more design consciousness than ourselves. Indeed it could be argued equally well that design awareness is in fact rising with affluence and educational attainment so that we are generally more aware of design quality, in clothing, cars, consumer goods, household artifacts, home decoration and even urban design than ever before. Perhaps that is why we know that a lot of it is not very good.

Secondly, there is always a long time lag between the development of a theory and its common acceptance and implementation, or between policy implementation and its widespread effects. Thus it is not suprising to find that the conservation-based view of urban design that has been accepted since the late sixties is only slowly beginning to have an impact. It takes time for society to acknowledge and reject the consequences of the collection of design decisions taken during the fifties and early sixties; it takes a long time and it is very difficult to train sufficient designers to move from the new urban design as experiment to the new urban design as normal practice. It is equally long term and difficult to train the planners, and the controllers, and to give them sufficient powers. In addition, there will always be a reactionary element, whether developer, architect or planning committee, that will make design decisions against the will of the majority. But progress has been made. One only has to look at the changing quality of decision making in the historic parts of any British town or city to see the growing sophistication and contextual awareness of urban design and planning decisions.

Thirdly, it has to be understood that design decisions are taken within an economic context and that the changing nature of development decisions inevitably put strains upon traditional scales and methods of development, use of traditional materials, styles and so forth. Sharp (1968) mourned the loss of perspective in Oxford's main street as national chain stores destroyed the series of individual shops which gave such a strong vertical emphasis to perspective views of Cornmarket Street, and replaced them with long buildings with a horizontal emphasis, out of rhythm with the rest of the street. But the decisions of these chain stores were dictated by the changing economics of retailing: large single-storey selling floors were necessary for competition and to their survival as firms. The hypermarket can be a beautifully designed box and it is the obvious economic solution to a particular retailing question but it poses the urban designer with near insoluble problems. It is frequently difficult to get developers to build sufficient upper storeys in shopping centres to provide the sense of enclosure that designers may be seeking but again urban economics dictate that there is limited demand for upper floors above shops; people do not like to live in them and institutional funders are cautious about investing in office developments above shops. From the discussion of economics in Chapter 3 it is clear that city centre office development is increasingly not a bespoke operation for some local enterprise but a standardised commodity to be occupied and traded easily on the property market. There is no room here for local foibles: conformity is the key to financial success.

7 Current Practice in Urban Renewal

This chapter deals with current policies in the urban renewal field and their implementation. It is structured around the three possible forms of state intervention: (i) the making and enforcing of laws and regulations; (ii) government spending and taxation policies; and (iii) organisational changes including the establishment of new agencies to deal with particular problems or policy approaches. In each section various policy instruments are described and critically discussed. Where appropriate case studies and examples are presented in order to illustrate and clarify the issues being discussed.

7.1 Intervention Through Regulation

Land use planning

In the reconstruction period after World War II and during the early years of slum clearance and town centre expansion in the 1950s the newly created statutory town planning system played a significant role in determining land use allocations and shaping the pattern of urban renewal in many cities. Under the 1947 Town and Country Planning Act all County Councils and County Boroughs were required to prepare Development Plans to indicate the way in which areas would be developed or redeveloped. Urban renewal was dealt with by Town Maps, which would be sufficiently detailed to show land use zoning and the density of use permitted, and Comprehensive Development Area Maps where substantial redevelopment was imminent. These latter maps would include details of street layout and sometimes the positioning of buildings.

For a variety of reasons these plans had fallen into some disrepute by the early sixties and in a review of the planning system in 1965 the Planning Advisory Group proposed a new two tier system with Structure Plans to deal with strategic issues of urban policy, and Local Plans to deal with the details of land use planning and development control (MHLG, 1965). The new system came into operation with the Town and Country Planning Act

1968. At that time, as shown by the Ministry's Development Plans Manual (MHLG, 1970), it was envisaged that Structure Plans would, *inter alia*, provide the broad context for urban renewal: identifying areas to be conserved and areas of change, indicating the general location of important transportation routes and major development proposals, and establishing broad policy objectives. Within this context District Plans (a type of Local Plan) would give a precise indication of locations and land use zoning, while, in areas subject to 'intensive change by development, redevelopment or improvement' Action Area Plans (a replacement for the former CDA maps) would provide detailed and precise guidance on building use, location, access, environmental works and so forth. Scarcely had this legislation received Royal Assent before it was being undermined by a whole series of events and alternative procedures.

Under the 1967 Civic Amenities Act local authorities were given the power to designate Conservation Areas within which they could work to preserve and enhance local character. This offered authorities a much faster mechanism for establishing their intentions in an area than waiting for the preparation and approval of a Structure Plan and subsequent Local Plan. Under the 1969 Housing Act local authorities were able to designate General Improvement Areas (GIAs) where individual dwellings could be identified for improvement or demolition and where detailed traffic management and environmental enhancement proposals could be put forward.

The concept of the GIA almost exactly mirrors the example of an Action Area Plan for a housing renewal area contained in the Ministry's Development Plans Manual but, again, offered a simpler, faster procedure. Within the major conurbations Passenger Transport Authorities were established in 1968 and charged with responsibility for public transport within their areas. As part of this duty they were required to prepare plans for the coordinated development of public transport provision. While they were obliged to have regard to the Structure Plan, they were, in effect, taking over an important element of town planning policy, transportation: a topic that had been effectively excluded from the 1947 planning system but was included within the 1968 system after critics had emphasised the close connection and interaction between land use activity and transportation investment decisions.

By far the most important event to undermine the planning system was the reorganisation of local government under the 1972 Local Government Act. Under this Act the new County Councils became responsible for the preparation of Structure Plans while most local planning and development control fell to District Councils. The problem was particularly acute in the metropolitan areas where, in most cases, the County Council had to practically restart the long Structure Plan preparation process. This exacerbated an already unsatisfactory situation: many cities were still using Development Plans prepared ten or fifteen years earlier for development control. Further, what had been one Development Plan preparation process was now split between two groups of planners (County and District), with different attitudes, approaches and abilities, and two groups of politicians with differing ideologies and objectives. By 1978 the situation had become so intolerable that the Inner Urban Areas Act permitted the adoption of Local Plans in inner city areas, under certain conditions, in advance of the Structure Plan. This change was indicative of the government's frustration with Structure Plans and its recognition of the near impossibility and limited value of making strategic urban policy statements in this form.

Under the 1980 Local Government, Planning and Land Act the government extended the provision for Local Plans to be adopted ahead of Structure Plan approval and with the abolition of the Metropolitan Counties in 1986 Metropolitan Structure Plans were to be replaced in due course by Unitary Development Plans to be prepared by Metropolitan District Councils within strategic guidelines to be established by regional offices of the Department of the Environment.

The statutory planning system has also been undermined by the local authorities themselves who, during the 1970s, increasingly began to use informal plans as the basis for decision making, and by developers (including local authorities) who took to using the development control system as a means of establishing approval for proposals for quite large areas.

Many authorities faced with declining inner city areas have taken the view that: (i) statutory development plans and the development control process have little to offer in the urban regeneration situation; and (ii) land use zoning was both difficult and of limited value in these areas where considerable change was taking place and many types of land use might be acceptable. In these cases it seemed more appropriate to rely on a 'criteria based approach', using such mechanisms as parking standards, noise and other pollution limits, and design criteria to control development. In other words for some authorities the approach to control in urban renewal areas tended to shift from the 'optimising' approach of rigid land use zoning to a 'satisficing' approach based upon development criteria.

This is not to say that the statutory planning system is no longer relevant or important in urban renewal, for indeed it is. In spite of some recent relaxations in the planning regime (in Urban Development Corporation Designated Areas, Enterprise Zones and Simplified Planning Zones, expansions in the amount of permitted development under the General Development Order and changes in the Use Classes Order (for example creation of the 'Business Use' class)) the fact that planning permission is required for most development still gives local planning authorities considerable power and influence over the urban renewal process.

In determining applications for planning permission authorities are required to 'have regard to the provisions of the development plan' and may also take into account 'any other material considerations'. In this way statutory Local Plans can be influential in determining the general nature and disposition of urban renewal and they offer the local authority a powerful weapon in the defence of a development control decision at any appeal. Through this local planning system the authority may:

- specify the location of particular land uses or individual developments;
- specify the intensity (density) of land use;
- identify changes in the road system, including new routes, and traffic management measures, including pedestrianisation of streets;
- require certain access and parking provisions to be met;
- designate areas for redevelopment, conservation or improvement and define the boundaries of other policy instruments such as Conservation Areas;
- specify design policies such as heights of buildings, protection of vistas, building lines, design styles and materials. The degree to which such policies are enforceable depends upon the nature of the area;
- protect existing buildings, landscapes and trees and propose landscape enhancement policies.

Because local authorities have discretion in the exercise of these powers and because the obtaining of planning permission is often a prerequisite to profitable development, the local authority can, particularly in a buoyant market, find itself in a good negotiating position *vis-à-vis* the developer. In this situation some authorities have been able to use 'Section 52 Agreements' to bargain for so-called 'planning gain' from developers. (S52 of the Town and Country Planning Act 1971 permits a local authority to enter into an agreement with a landowner for the purpose of restricting or regulating development beyond their normal powers of planning control.)

Such gains take a variety of forms but typical examples would include provision of land for low cost housing, additional landscaping or access agreements, provision of additional car parking or space for social and community facilities. The extraction of 'planning gain' leads local authorities into a potentially shady area of legality and ethics and in recent years the government has sought to tighten controls on what constitutes acceptable practice.

The major shortcomings of limitations on the role of the planning system in urban renewal are that:

(i) it can only react to proposals arising from developers and can do little to encourage or stimulate development activity in times or areas of low demand;

(ii) it cannot prevent the cessation of activity on land (i.e. it cannot prevent the cessation of employment or habitation) and only in a few specific instances, such as Listed Buildings, can the authority prevent building demolition;

(iii) it can do little to control who occupies land or buildings or how these are used within the limits of the use classes.

Several Structure Plans prepared during the 1970s sought to encourage urban regeneration through placing restrictions on peripheral development in the hope of forcing development back into the inner city. The Merseyside Structure Plan was built around such a policy:

> The strategy aims to concentrate investment in the built-up areas of Merseyside, especially those which have the worst problems of decline. It will improve the environment and encourage development on derelict and disused land. The strategy makes a parallel effort to protect and improve the countryside. It will restrain building on the outskirts of towns, encourage efficient farming and make proper use of the countryside's potential for leisure and sport.
>
> (Merseyside County Council, 1979, p.8)

Given the problems facing Merseyside it is difficult to see any alternative to such a strategy but equally, given the historical trends towards suburbanisation and regional decline it is difficult to see how such a strategy could do more than ameliorate an inevitable long run decline of the conurbation and gradual reduction in the intensity of occupation and use of the inner city. Indeed in the ten years since the preparation of the plan while building has been very effectively restrained on the outskirts of towns, there is very little evidence to suggest that much of this has been diverted towards the inner city. Admittedly there has been an upturn in private investment in the older urban areas but much of this is the result of substantial government subsidy or city centre commercial development and has been little influenced by land use controls at the periphery.

Compared with most other public agencies local planning authorities have had a good record of public consultation and some public participation in the plan making process but the record is let down as some plans have become discredited and public expectations have not been realised. Beyond plan making in the realms of plan implementation and development control, planning authorities have generally reverted to a more closed decision-making process. Again, compared with many branches of central and local government planning departments do tend to show significant adaptive–organic organisational characteristics but few have developed the open and participative style of decision making that could really enhance their role in the urban renewal process.

Liverpool City Centre Strategy Review

Although not a formal statutory Local Plan the Liverpool City Centre Strategy Review provides an interesting example of how a local authority is trying to get to grips with the problem of controlling the urban renewal process in a city centre and to manipulate events towards achieving wider public objectives through a comprehensive overall plan or policy document.

In 1988 Liverpool City Planning Department prepared a document reviewing the, by then, twenty-year-old city centre plan. The original plan for city centre development and renewal had been prepared in an era of optimism about future prospects for the growth of employment and investment and an anticipation of large scale public sector infrastructure construction. Since that time it has become apparent that the levels of retailing and office employment and investment in the city centre are far less than was hoped for at that time, and that changes in policy and levels of public spending have curtailed former ambitions for high level pedestrian routeways, grade separated highway systems and a new civic centre. It was in this climate that the new strategy review was prepared.

The document is interesting as an urban renewal policy since it is seeking to provide a comprehensive framework for the renewal of a very complex area with many economic and social forces pulling in different directions and a number of potentially conflicting policy objectives. The review considers the needs of the city centre under a number of themes and then establishes priorities for action. The themes are as follows.

(i) **Responding to changing retail market trends**. Liverpool is a major regional retail centre but has stagnated in recent years and is facing strong competition from competing centres such as Chester, Warrington and Southport. The proposed response is to recognise the importance of the city centre in the local economy and to call upon formal Development Plan policies to safeguard this role, for example by restricting peripheral out-of-town shopping developments.

(ii) **Strengthening links between Liverpool and its catchment**. The response is to argue in a rather unfocussed way for more highway investment, an end to Mersey Tunnel tolls, and more central government financial support for bus and rail public transport systems.

(iii) **Improving the circulatory system for pedestrians and vehicles**. The present vehicular circulation system is confusing. Car parking is inadequate and insecure; there are several points of severe pedestrian/vehicular conflict and bus termini are badly sited. The response is to call for a reappraisal of the internal circulation system within the city centre to create a 'visitor and user friendly' system.

(iv) **Upgrading the environment**. The city centre has a renowned architectural and historic heritage and a superb physical setting which has in some cases been undermined by blight, poor development control and low levels of environmental maintenance (e.g. refuse collection, street cleaning). The document argues for stronger development control policies and more spending on basic environmental services.

(v) **Developing tourism**. Recent years have seen a major growth in tourism in the city but further growth is said to be inhibited by a poor image and specific problems of environment, parking and circulation. The response is to call for a 'Tourism Plan' comprising proposals for physical developments, enhancement of major attractions, appraisal of visitor services, marketing, and tourism employment and training policies.

(vi) **Changing commercial requirements (offices)**. There is a problem of a historic office core of attractive exteriors and inefficient interiors, large scale losses in office employment and a lack of new investment. The response has been to establish a Town Scheme in the conservation area, to study the refurbishment potential of the existing stock, to identify needs for new stock and to market the locational advantages of the city.

(vii) **Increasing housing opportunities**. In an attempt to get more people living back in the city centre as part of the city's wider urban regeneration strategy two or three housing areas adjoining the city centre have been the subject of intensive upgrading works.

(viii) **Reinforcing the city centre's artistic and cultural importance**. There is a perceived need for more sponsorship and the attraction of bigger audiences to top quality entertainment. Again, in a similar way to tourism policy, a comprehensive package is proposed including appropriate land use zoning, marketing of the arts, the promotion of new activity, arts employment and training policies and community involvement and access.

(ix) **Promoting a positive image for the area**. It is felt that the area has suffered from various forms of bad publicity which in turn have had an adverse effect upon investment. Thus part of the urban renewal policy is to recognise and react to this situation.

From this discussion of themes four immediate priorities for action are then established: the preparation of a Unitary Development Plan (a single plan that replaces the former Structure and Local Plan system in metropolitan areas); a thorough appraisal of the internal circulation and parking system within the city centre; immediate improvements to environmental maintenance and security programmes; preparation of a marketing strategy for the city centre.

Listed buildings and Conservation Areas

For many years the Department of the Environment has taken responsibility for the 'listing' and protection of buildings and other structures of architectural or historic interest. Currently virtually all pre-eighteenth-century buildings and most buildings up to 1840 are listed. Beyond that a considerable number of buildings of quality and character built before 1914 and a small number of more recent buildings are listed. Currently there are well over 250,000 such buildings in England and Wales alone.

Listed buildings are graded as I, the most important buildings that should be afforded maximum protection and preservation effort, and as grade II, those of less importance. The purposes of listing are (i) to provide guidance to local authorities and others in the detailed physical planning of areas and in the provision of subsidies to assist the repair and improvement of buildings, and (ii) to ensure that no listed building is demolished or materially altered without 'listed building consent'.

Applications for listed building consent must be advertised and will be considered by the local authority after consulting interested amenity and historical societies. After this the authority may refer the matter to the Secretary of State who may vary the decision of the local authority. There is a general assumption against demolition of listed buildings.

It was not until the introduction of Conservation Areas in the Civic Amenities Act 1967 that there was statutory recognition for the idea that:

> the character and historic value of a town or village often depends as much on areas as on individual buildings
>
> (RICS, 1983, p.9)

and determined that:

> local authorities have a duty to determine which parts of their area are of special architectural or historic interest, the character of which it is desirable to preserve or enhance as Conservation Areas.
>
> (RICS, 1983, p.9)

There are currently over 5,000 Conservation Areas in England and Wales. In Liverpool, for example, the designated areas include such diverse environments as the historic core of the city centre, Georgian terraces, former village centres and even some planned council housing estates from the inter-war period. Figure 7.1 illustrates this variety.

Within a Conservation Area grants may be available from the Department of the Environment to assist with repairs and improvements to property. These grants, usually obtained under the Town and Country Planning (Amendment) Act 1972 may be for up to 25 per cent or 50 per cent of the costs incurred. Local authorities can also make grants or loans towards the cost of repairing buildings of architectural or historic interest.

In order to enhance the character and quality of a whole Conservation Area, including giving help to improve unlisted buildings, a local authority can declare a 'Town Scheme'.

Figure 7.1 Conservation Areas in Liverpool

> A Town Scheme provides grants for structural work to selected buildings that may not be individually outstanding, but contribute to the character of a conservation area. Usually the DOE and local authorities agree to allocate fixed sums for a five year programme period. Grants are for works of structural repair at a level of 50 per cent.
>
> (Thomas, 1983, p.254)

Focussing upon the conservation of residential areas, Thomas (1983) has identified a number of problems and conflicts that are likely to arise in the implementation of Conservation Areas. He summarises these in a table which is reproduced as Table 7.1.

The designation of a Conservation Area gives the local authority additional planning controls, in particular the withdrawal of rights of 'permitted development' under the General Development Order (which would normally allow minor building extensions and alterations) by means of an 'Article 4 Direction'.This procedure gives the authority additional controls over design, including materials and colour schemes and the use of buildings.

Recently some assessment of the effectiveness of Conservation Area policy has been carried out at Liverpool Polytechnic (Morgan and Nixon, 1988). The study sought to obtain and analyse the views of planners, surveyors, architects, amenity societies and the public in a sample of Conservation Areas across England. From this work Morgan and Nixon make a number of comments on the success of Conservation Areas and suggest some scope for policy development.

They found agreement amongst most of the planners consulted that the current Conservation Area controls had helped to improve the areas and that the general building fabric was now in better condition than might otherwise have been the case because planners were able to enforce certain types of maintenance work upon owners. Many of the surveyors and agents questioned remarked upon the higher costs of developments in Conservation Areas but equally all acknowledged that there was a general tendency for a higher level of demand for property within such areas and a consequent rise in values compared with equivalent areas nearby. Thus increased costs were being passed on to occupiers or consumers and development profits maintained.

In general the authors felt that:

> there exists a great deal of stereotyping with, for example, distrust between surveyors and amenity societies based upon what they think the other party would wish to do and not on what is actually the case. A forum needs to be established to facilitate a positive exchange of views

Table 7.1 Conflicts in conservation policy

<table>
<tr><td>Local needs
Improved housing conditions for local people at reasonable cost.</td><td>National interest
Preservation of the national heritage to a high standard of historical integrity and with a view to the economic advantages of tourism.</td></tr>
<tr><td colspan="2">This conflict between local needs and national interests is, therefore, broadly a conflict of costs and standards; but it also involves the issue of use.</td></tr>
<tr><td>Property investment
Capital investment is essential for the renovation and maintenance of older property.</td><td>Local housing requirements
With investment comes the danger of increased property values, so pricing local residents out of the housing market.</td></tr>
<tr><td colspan="2">Here the argument is whether the maximum investment consistent with maintaining existing social use is adequate to maintain the building.</td></tr>
<tr><td>Tourism
Tourism development brings local economic growth, increased service provision and property maintenance.</td><td>Local residents
Tourist traffic brings pollution, visual intrusion and increased congestion.</td></tr>
<tr><td colspan="2">Tourism promises to generate the wealth necessary for investment in maintaining the physical environment. But even if this is achieved, it may be at a cost to the quality of life in the local community.</td></tr>
<tr><td>Preservation
Retaining structural and historic integrity.</td><td>Utility and function
Adapting for twentieth-century needs with restricted resources.</td></tr>
<tr><td colspan="2">This is the conflict between preservation and conservation. If buildings are to be retained by adapting them to modern requirements, there will be a set of conflicts in terms of standards, function, convenience and preservation.</td></tr>
<tr><td>Townscape
Residential property may contribute to the essential character of the built environment.</td><td>Social and environmental obsolescence
The same property may be surplus to local requirements or be located in an undesirable environment.</td></tr>
<tr><td colspan="2">In essence, therefore, buildings may perform an important visual task, while lacking any functional value.</td></tr>
</table>

Source: From Thomas A 'Planning in Residential Conservation Areas' *Progress in Planning* Vol. 20 p. 233.

by all interested parties, placing a new emphasis on pro-active participation.

(Morgan and Nixon, 1988)

An important point that has long been recognised but which was brought out again in this study was that there could frequently be a conflict of interest in such areas between the local authority as landowner or promoter of economic development on the one hand, and the local authority as guardian of amenity and character in areas of architectural and historic interest, on the other. The general view of architects, amenity groups and others was that design control should always be enforced, but given this real conflict and the number of infringments and concessions to short-term expediency that are constantly being reported in the planning press it is worth raising the question as to whether such control should be transferred into the hands of some separate public agency. Indeed this raises a much wider question about the compatibility of local authorities' multiplicity of traditional regulatory and enforcement duties (in land use planning, public health, trading standards, etc.) and their more recent acquisition of promotional roles (in economic development, tourism, etc).

It raises even more difficult questions about the role of Urban Development Corporations which have both planning and development powers and are not subject to any form of local accountability whatsoever. It might be argued that it would be in the long run interests of 'better' urban renewal if there were to be a separation of these conflicting powers and duties and, especially in view of growing ecological and environmental concern, a more rigorous enforcement of local health regulations, land use zoning, design and environmental standards.

Land registers

During the late seventies concern was rising about the amount of vacant and derelict land within urban areas. The phenomenon was investigated in a number of studies including those by the DOE (1975b) and Nabarro (1979). There was also considerable debate in the property press which tended to emphasise, rightly or wrongly, the amount of public sector owned land that was being withheld from development for highway schemes that had been abandoned, housing schemes that could no longer be afforded, or in anticipation of market prices that were unrealistically high.

Within this context the government, under the 1980 Local Government Planning and Land Act required all local authorities to provide the Department of the Environment with information on all surplus unused and underused land owned by public bodies. These registers are used to bring the existence of such land to the attention of potential purchasers and developers. The Secretary of State can direct the disposal of sites if he thinks that it is being unreasonably withheld by the public sector.

To some extent this is a solution to a problem that is already on the wane. Much of the land in question became vacant for the reasons outlined above: the abandonment of road schemes, slum clearance programmes moving too far ahead of replacement building, rationalisation and closure of railway marshalling yards, closure of town gas works, etc. Many of these were one-off events that happened to coincide in time in the early seventies and are unlikely ever to be repeated on such a scale. Gradually both public and private sector development programmes are consuming this backlog of vacant public sector land. This is not to say that dereliction is becoming a less serious problem, but it is increasingly shifting to the private sector as industrial and commercial premises are abandoned and not re-used. This is discussed again later under 'Derelict Land Grants'.

A further point is that as the financial crisis of inner city local authorities has worsened the majority have been only too keen to dispose of surplus land, seeing the capital receipts as a means of supplementing their meagre resources.

7.2 Intervention through Spending and Taxation

Derelict Land Grant

According to the Department of the Environment's own survey there were 45,683 hectares of derelict land in England in 1982 (DOE, 1984). That was 'land so damaged by industrial or other development that it is incapable of beneficial use without treatment'. Most policymakers and analysts, including the DOE itself, accept that this is an underestimate since small sites (of less than 0.4 hectares) and other vacant sites not falling within this definition are excluded. Concern over dereliction and particularly the problem of 'vacant urban land' has developed over the past 25 years and generated considerable debate and political action. The reasons for dereliction and vacancy have been discussed earlier in Chapter 3 and at length elsewhere (see: Cameron *et al.*, 1988, and Couch *et al.*, 1989).

There are a number of policies used to tackle urban vacancy and dereliction, including use of the Registers of Vacant Public Sector Land, spending through the Urban Programme and grant assistance under Industrial and Commercial Improvement Areas policy. However, by far the most important instrument over many years has been Derelict Land Grant (DLG).

Since the mid-sixties the government has offered Derelict Land Grant to local authorities to facilitate the reclamation of land falling within the statutory definition of dereliction. This has been a major spending programme which has grown in recent years from an annual expenditure of about £21m in 1979 to about £81m in 1987/88. Until 1981 only local authorities were eligible to receive DLG but since that time the grant has also been available to private companies, nationalised industries, charitable trusts and others.

> In Assisted Areas and Derelict Land Clearance Areas grant is at the rate of 100 per cent for local authorities and English Estates and 80 per cent for others. The rate elsewhere is usually 50 per cent. The amount of DLG paid is the appropriate rate on any net loss incurred in carrying out an approved reclamation scheme. The allowable cost is generally what is needed to bring the land to the equivalent of a 'greenfield' site....Priority for funding is given to reclaiming derelict sites in urban, especially inner city, areas to provide land for private sector housing, industrial and commercial development....Since 1979/80 over 10,600 hectares of derelict land have been reclaimed in England with the aid of DLG...Over this period, the proportion of reclaimed land intended for subsequent development (rather than improved for purely environmental or open space purposes) has increased from 6 per cent to 50 per cent and over a similar period DLG expenditure in inner city areas has risen from 7 per cent to 30 per cent of the total.
>
> (DOE, 1988, p.45)

There have been two separate recent official evaluations of the effectiveness and efficiency of DLG. In 1987 a DOE sponsored study by Tym and Partners was published, to be followed in 1988 by a report from the National Audit Office.

The DOE has four main objectives for reclamation including: the removal of danger and improving environmental quality; the provision of open space and land for recreation; the provision of land for employment-producing uses and housing; and, encouraging investment in adjoining areas as environmental upgrading improves business and investor confidence.

The findings of Tym *et al.* suggests that the environmental and safety objectives are being met in virtually all schemes but that objectives relating to the provisions of development land have proved more difficult to achieve, or at least are taking longer. About two-fifths of the sites intended for development and treated more than four years before the publication of the report were still vacant. It was also noted that such re-use is harder to achieve in some locations than others. However, there did seem to be some benefit to confidence in surrounding areas from reclamation and environmental improvements particularly if a large area or group of sites had been tackled.

From their conclusions, Tym *et al.* recommended that the increased emphasis on 'hard' end uses (i.e. money making development) should be applied flexibly depending upon the market conditions in the area. They also favoured the development of rolling programmes to increase efficiency within local reclamation agencies and to give emphasis to concentrations of effort to obtain the 'critical mass' necessary to generate changes in business perceptions of an area. Where schemes were for 'soft' end uses local agencies were urged to be much more imaginative in the design of open space and 'in creating richer opportunities for active and passive recreation' (Tym *et al.*, 1987, p.52).

While the National Audit Office took a narrower view, being concerned mainly with departmental efficiency, it concluded that:

> The reclamation of derelict land is pursued by the Department's regional offices and by local authorities with vigour and determination to make the best use of the funds available; and the programme has over the years produced substantial benefits and major individual successes. Nevertheless there remain important doubts on effectiveness and value for money, and more needs to be done to strengthen monitoring procedures, to improve analysis and follow up of results, and to introduce satisfactory management information and control systems.
>
> (National Audit Office, 1988a, p.3)

The report by Tym *et al.* (1987) contains a number of case studies of the successful implementation of derelict land reclamation schemes. One example of good practice is summarised here

Widnes in Cheshire is one of the traditional centres of the country's chemical industry. As the industry has rationalised in recent years the town has been left with a legacy of large tracts of contaminated land unsuitable for re-use without substantial expenditure on reclamation and treatment works. Since 1974 Halton Borough Council, the local authority, in conjunction with Cheshire County Council have been responsible for the reclamation of over 117 hectares of derelict land. One of the policies adopted by the local authority was to concentrate reclamation work along corridors that could be seen from major communication routes so as to improve the image of the town, principally to attract inward investment. One such corridor was known as Ashley Way, an inner relief road around the south of the town centre. Along the route of this road four separate major sites were reclaimed between 1982 and 1986. These were:

(i) St Maries, a 0.9 hectare cleared housing site, owned by Halton Borough Council, it cost £28,052 to reclaim and receive 100 per cent DLG. In addition Halton Borough contributed a further £15,925 for environmental enhancement. The after-use is for public open space associated with adjoining housing.

(ii) Dock Junction, a 2.4 hectare disused railway line, purchased by

Halton Borough. Acquisition and reclamation costs of £51,647 attracted 100 per cent DLG. Again the Borough contributed a further £14,945 in enhancement costs to provide for public open space after-use.

(iii) Ann Street, 16.2 hectares of contaminated chemical land and redundant railway line. Purchased by Cheshire County Council and reclaimed in three phases at a total estimated cost of around £2.3 million, with 100 per cent DLG. The intended after-use is for commercial and industrial development.

(iv) Vine Chemicals, 8.5 hectares of former chemical works, acquired by Halton Borough at a cost of £718,750 with 100 per cent DLG. Reclamation costs are not yet known but the intended after-use is town centre expansion.

In addition to DLG three other major sources of funding have been used in environmental improvement work: the Urban Programme for a 'facelift' scheme; a second urban programme project known as Halton Employment Action Team used for building renovation; and the Borough's own capital works programme paid for certain landscaping works. In addition a minor contribution came from a Section 52 Agreement with a private housing developer (planning gain).

Implementation of these reclamation schemes was piecemeal and incremental, with the local authority taking opportunities as they arose rather than working to a rigid pre-conceived timetable. The effect of the programme has been that:

> The improvement works along Ashley Way have created a pleasant green corridor. Design, implementation and management of the landscape works have served to break down Widnes's 'dirty' image.
>
> (Tym *et al.*, 1987, p. 70)

Most businesses in the area considered the environmental improvements to have been of value in attracting business activity and investment to the area. The Chamber of Commerce also felt that business confidence had improved and there has been a significant recent increase in land values in the town, although it was difficult to relate these trends to any one specific cause.

The Urban Programme

The Urban Programme began life as 'Urban Aid' in the late 1960s as a Home Office programme of assistance to inner city community and environmental projects. Its importance grew through the 1970s until the passing of the 1978 Inner Urban Areas Act when it emerged as the 'Urban Programme' firmly under the wing of the Department of the

Environment which, at that time, had become the lead department on inner city matters.

The modern Urban Programme is one of the most important elements of direct government financial support for inner city projects. Currently DOE spending on the programme, at £221m per annum, is equal to the amount allocated for Urban Development Corporations. However, the Urban Programme also includes special allocations from other departments (DES, £44m, 1988/89; DHSS, £38m, 1988/89; DTp, £11m, 1988/ 89) all co-ordinated by the DOE.

The aims of the Urban Programme are to achieve cooperation and coordination between relevant agencies, the private sector and the local community within the 57 inner city local authorities (the 9 Partnership and 48 Programme Authorities). Expenditure occurs under two main elements: (i) the Inner Area Programmes prepared annually by these local authorities; (ii) City Grant (formerly Urban Development Grant and Urban Regeneration Grant) under which private sector projects may bid for government assistance (see below).

An impression of the nature of Inner Area Programme spending can be derived from a look at the achievements of the 1987/88 programme in Table 7.2. Expenditure is classified under four objectives: economic; environmental; social, and housing.

In a recent assessment of environmental projects carried out with Urban Programme funding the consultants, JURUE, concluded that success was susceptible to a large number of factors. These included:

- Location and context. Improvements have been most effective in prominent locations, higher density locations and where the private sector has been willing to invest.
- Involvement of the voluntary sector was seen to be particularly useful.
- Environmental programmes involving many similar projects have been helpful in generating local experience and improving the performance of individual projects.
- In the management of projects the close involvement of beneficiaries, the undertaking of projects within clearly defined objectives, giving due consideration and attention to after care and maintenance are important elements.

(JURUE, 1986b)

Urban Development Grants (UDG) were launched in 1982 with the aim of promoting economic and physical regeneration in run down urban areas by encouraging private investment that would not otherwise take place. The purpose of UDG was to bridge the gap between the cost of a development and its value on completion. Assistance could be in the form of grants or loans for any size of project. Applications had to be made

Table 7.2 Inner Area Programme: 1987/88 outputs

	£ million	Outputs
Economic objectives:		
(i) new factories, enterprise workshop/commercial units	10.2	1,020 units; 4,220 jobs created/ preserved
(ii) buildings improved/converted for factories/enterprise workshop/commercial units	14.1	860 units; 2,670 jobs
(iii) business starts in enterprise workshops (included above)	2.2	560 business starts
(iv) training places supported	17.8	c.60,000 places
Environmental objectives:		
(i) buildings improved	7.8	780 shops; 1,230 other buildings
(ii) land improved	27.3	750ha. derelict; vacant land; 1,180ha. parks and open space
(iii) routes improved	6.8	130km waterways; 210km roads/paths
Social objectives:		
(i) community centres	19.3	530 projects
(ii) sport/recreation projects	25.1	2,590 projects
(iii) health projects	13.1	419 projects
Housing Objectives:		
(i) accommodation in refuges/ hostels	4.3	160 schemes; 4,250 bedspaces
(ii) environmental improvements in problem estates	8.8	56,000 dwellings benefitted
(iii) management improvement schemes	2.4	89,000 dwellings benefitted

Source: H M Treasury, 1988 *Government Expenditure Plans 1988–89 to 1990–91*, HMSO, London.

through local authorities. Urban Regeneration Grant was launched in 1987 to complement UDG by enabling the private sector to redevelop large sites and refurbish large groups of buildings. It differed from UDG in that it was directed at larger scale projects, the longer term commitment of resources and did not need local authority involvement (DOE, 1988, p.47). Both types of grant were absorbed into the new City Grant under the government's 'Action For Cities' programme in 1988.

With regard to the effectiveness of UDG it has been suggested that:

> The encouragement of private developers has given undue preference to commercially viable proposals at the expense of socially worthwhile projects. Nevertheless, there can be no denying the generative impact of many of the large schemes which have been set in train. By 1987, some £130m of Government funds had been allocated to well over 200 projects and these had attracted an additional £520m of private sector funds, thereby producing an overall gearing ratio of 1:4.
>
> (Robson, 1988, p. 118)

Enterprise Zones

In the 1980 budget the Chancellor announced the experimental concept of Enterprise Zones (EZ)within which certain administrative controls would be relaxed and financial burdens lifted in an attempt to encourage additional industrial and commercial activity. Whether this is the right place to consider a policy that was part regulation and part spending (in the sense of taxes foregone) is a moot point, but we consider it here for convenience.

Enterprise Zones are designated for ten years. Within the zones and for the duration of designation, firms are exempted from local rates (which central government reimburses to the local authority), receive 100 per cent allowances against Corporation Tax or Income Tax for capital expenditure on buildings, are subject to a simplified town planning regime with automatic permission for many types of development, and further minor concessions.

By 1983 there were 23 Enterprise Zones in Great Britain ranging from about 50 hectares to 452 hectares (although the latter was on a number of sites) and while the majority are to be found in peripheral regions: Central Scotland, the North and North East, the North West and South Wales; a small number have been located in areas that were perceived as pockets of depression in otherwise prosperous regions, e.g. the Isle of Dogs (London), Corby, Wellingborough.

In the early years of implementation the Department of the Environment sponsored a series of annual monitoring reports from Tym and Partners, the final one appearing in 1984. This and other studies (e.g. Erickson and Syms,1986) suggested that while investment had evidently been attracted in varying degrees to different Enterprise Zones and jobs had resulted, much of this would have occurred anyway and was merely being transferred into the zone from elsewhere in the region. It was also noted that land prices within the zones were being bid upwards while the price of neighbouring land was reducing, thus a proportion of the financial benefit to firms of being inside a zone was being lost to the land market.

Nevertheless there have been some spectacular developments within some zones. The Isle of Dogs Zone was declared at a time when the London Docklands Development Corporation had just been designated and when it was by no means certain that there was a significant demand for land that far east of the City of London. The EZ was envisaged as a spur to industrial development. As it turned out the zone coincided with a boom in demand for office floorspace in London and a boom in the property market. Early ideas of low density industry were quickly swept aside by massive investment in offices and other forms of intense commercial development culminating in the Canary Wharf proposal for 11 million square feet (over 1 million square metres) of office floorspace, currently under construction. In another example of local entrepreneurship and sheer audacity in the Gateshead EZ, a local developer piloted through the building of the biggest enclosed freestanding shopping centre in Europe: the Metro Centre. Whatever its merits as a centre and its long run impact on neighbouring towns, in terms of attracting investment to a most unlikely location it has to be seen as a success and an indication of what can be achieved through individual initiative when the tide of government policy and economic forces are working in the same direction.

During 1987 the Department of the Environment published a series of commissioned research papers reviewing various aspects of their inner area policies. A study of Enterprise Zones was carried out by PA Cambridge Economic Consultants (DOE, 1987). The report came to the following conclusions.

- In providing a balance of supply side and demand side stimuli the experiment had been well designed.
- The experiment had attracted private finance capital into areas where it would not otherwise have penetrated.
- Of 63,300 jobs within EZs, it was estimated that about 35,000 had been created as a direct consequence of EZ policy although many were merely transfers from elsewhere in the local economy, however.
- Because of local multiplier effects 'about 13,000 net additional jobs are supported directly and indirectly in the local economies by the EZ experiment' (DOE, 1977, p.86).
- About 23 per cent of firms in EZs were already there, 14 per cent are new branch plants, 37 per cent are firms transferring into the zone, but 'perhaps of most significance in the long run is the stimulation

given to the start-up of new independent small businesses (26 per cent)' (DOE, 1977, p.86).

- 'Another benefit generated by Enterprise Zones is the removal of physical dereliction and the improvement in the local environment. This environmental improvement will contribute to the creation of further new jobs far into the future. Modern firms are attracted to environmentally attractive areas and these benefits affect not only the designated area itself but other city centre areas adjacent to it.' (DOE, 1977, p.87).
- The general view was that exemption from rates was the most effective single incentive within the EZ package although a significant number of firms felt that they had benefitted from the more relaxed planning regime and/or capital allowances.
- Finally the report noted that while real benefits are being provided to local economies by EZs future adjustments to the policy might introduce some differentiation in benefits between zones so that subsidies would not be wasted in supporting developments that would be likely to occur anyway in the more prosperous areas (DOE, 1977, p.88).

Industrial and Commercial Improvement Areas

Following an initiative developed by Rochdale Borough Council in the mid-seventies, the Inner Urban Areas Act 1978 gave designated (partnership and programme) authorities power to declare Industrial or Commercial Improvement Areas. They could provide financial assistance to firms to help with the costs of environmental improvements: building walls or fences, landscaping, clearance and levelling of derelict sites and buildings, cleaning of building facades, improvements to access and parking arrangements (Section 5); and grants towards the cost of conversion, extension, improvement or modification of an industrial or commercial building (Section 6). The level of grant available was up to 50 per cent although the total amount available to local authorities was subject to cash limits.

The original aim of the policy was to stimulate job creation in these areas although it was recognised from the beginning that localised environmental improvements could, at best, have only a marginally beneficial effect on employment. In practice most authorities regarded the achievement of environmental improvements as a sufficient end in itself.

In 1986 an extensive evaluation of the effectiveness of the policy was carried out for the government (JURUE, 1986a). By the time the review was carried out some 212 IIAs/CIAs had been declared covering about 7,000 hectares of land: a considerable proportion of the industrial and commercial land within these inner city districts.

In the view of the consultants the measures pursued had increased business confidence in the area although, unlike the effect on housing in GIAs, this was only rarely matched by increased investment, and the effect on jobs was minimal. Indeed the main impacts of IIAs/CIAs had been to improve the visual appearance of areas and to achieve small scale, but important, physical changes (site clearances, access improvements, etc.) which would probably not have occurred without declaration. It was also found that there were actions, such as land assembly, that local authorities could have carried out under other legislation, that would have enhanced the success of IIAs/CIAs.

It was concluded that IIAs/CIAs were most likely to be successful in areas:

> 'where some changes stemming from market forces is likely to occur ... where physical improvements to premises are likely to visually benefit adjoining property, and where Improvement Area declarations coincide with changes in public planning policies towards the areas...'
>
> (JURUE, 1986, p.10)

It was also recognised that changing the industrial fortunes of an area is a slow process and that because of the likelihood of unfavourable short term market conditions greater emphasis should be given to long term prospects in the selection of areas, and then in turn a longer timespan be envisaged for the achievement of significant economic development (perhaps ten years instead of the current five recommended by the DOE).

Better monitoring and the preparation of 'action plans' in consultation with firms affected were also recommended.

One of the more successful examples of an Industrial Improvement Area can be found in the Lace Market area of Nottingham. The area lies to the east of Nottingham city centre and was for many years characterised by rundown industrial and marginal retailing and commercial uses but with a physical structure and built form that at least merited an attempt at conservation and enhancement.

In 1969 Nottingham City Council designated the Lace Market as a Conservation Area in order to prevent further demolition and the loss of character that was becoming an inevitable outcome of economic decline. In 1974 the area's status was upgraded to that of a 'Conservation Area of Outstanding National Importance' which allowed access to more funds for building repairs and refurbishment.

At this stage the City Council prepared a strategy for the future of the area intended to eliminate blight and encourage the restoration and re-use of buildings while taking account of the needs of local businesses, especially the still important textiles and clothing sectors.

In 1976 the City Council cooperated with the County Council and the Department of the Environment in establishing a 'Town Scheme' for the area and in 1978 still further assistance was available when the area was designated an Industrial Improvement Area. By this time some £1 million had been spent on improvement work, nevertheless a City Council report noted that the area was still run down and could not yet be said to have moved into a self-sustaining cycle of recovery. With this in mind the objectives of the Industrial Improvement Area were established as: the reversal of economic decline; the encouragement of private sector investment in environmental improvements; and, the consolidation of work already undertaken.

The actual improvements subsequently carried out included:

- the conversion of four council owned buildings for various uses, including offices, workshops and a community arts complex. Various other vacant buildings have been brought back into use;
- some 47 grants under Section 5 of the 1978 Act and six grants under Section 6 of the Act totalling £400,000 and generating at least an equivalent amount of private investments;
- a facelift scheme to improve street facades and the appearance of retailing areas;
- landscaping, car parks, open space and environmental works and some traffic management improvements.

A recent official publication on good practice in urban regeneration praises work in the Lace Market for its clear definition of the area, good use of existing features, careful consideration of priorities, the scale of the private sector contribution, the diversity of public funding, the strong visual impact of improvements and the functional improvements achieved. (JURUE, 1988, p.35 *et seq.*)

The re-use of urban land and buildings: some policy lessons from Germany and France

Both Germany and France also experience problems of urban vacancy and dereliction, although neither on the scale of Britain. From their experiences and approaches to policy some ideas and lessons emerge that might be of value in Britain.

In the Ruhr alone in Germany it has been estimated that there are over 2,000 hectares of derelict land. It was to tackle this problem that, in 1979, the government of North-Rhine Westphalia (NRW) set up the *Gründstückfond* (a fund for financing derelict land reclamation) to permit direct state intervention in a situation where market forces had clearly failed. Resorting to direct intervention in the land market is unusual in West Germany where all levels of government usually prefer to 'oil the wheels' of the market by means of tax incentives or subsidies rather than 'take over' some aspect of economic life. This demonstrates the seriousness with which the NRW government viewed the problem. The purpose of the

Gründstückfond is to assist economic development and to create environmental improvements through purchase, reclamation and resale of derelict land.

The actual management of land reclamation is delegated to the Landeswicklungsgesellschaft (LEG), a private limited company wholly owned by public agencies in North-Rhine Westphalia. The LEG is responsible for all negotiations with owners and purchasers, as well as the land reclamation itself. The government of NRW retains the right to approve buying and selling prices of land, and the local authorities in the area retain land use planning control over redevelopment. The local authorities also have the power to veto any sales of reclaimed land if they feel the purchaser is in some way inappropriate to the best interests of the locality.

Since its inception the system has brought about the reclamation of a quarter of the originally identified derelict land. While this is a reasonable level of performance it is no better than that which has been achieved in Britain in recent years. However three features of the policy are significant and might offer lessons to British policymakers. Firstly, the *Gründstückfond* is seen by the government of NRW as a long term investment programme. There is no requirement to achieve commercial (or indeed any particular) rates of return. This can be contrasted with the British government's attitude in which public development agencies are increasingly given commercial targets and are under strong pressure to accept any profitable development regardless of its long term value to the local community.

Secondly, as a regional programme the *Gründstückfond* LEG can take a wide spatial view of need and priorities for reclamation. In Britain equivalent agencies such as the Urban Development Corporations are restricted to working in a defined area regardless of the region's general needs and priorities. Thirdly, it has been the predilection of British governments to set up non-democratic, non-accountable organisations (New Town Development Corporations, Development Agencies, Urban Development Corporations) usually on the grounds that there is a need for speedy investment and efficient decision making. The *Gründstückfond*/LEG demonstrates how it is perfectly possible to achieve an efficient programme while retaining control within the democratically elected framework of local and regional government (Couch and Herson 1986).

In France also there are approaches to the re-use of land and buildings that are different from our own. While it must be stressed that the physical, economic and institutional context of urban renewal is very different in France the essential problem remains the same and it is still possible to consider the relevance of French policy to our own needs. Here brief details are presented on some of the more interesting French approaches to the planning, land assembly and implementation of urban renewal: the

Zone d'Aménagement Concerté; Zone d'Intervention Foncière; Zone d'Aménagement Différé: and the use of *Sociétés d'Economie Mixte*.

The *Zone d'Aménagement Concerté (ZAC)* is designed to achieve the comprehensive planning and development of an area of intensive change. Its closest British equivalent would be an Action Area Plan or the former Comprehensive Development Area (CDA). A *ZAC* is a zone where either a substantial amount of new development is anticipated (an industrial or housing estate) or where a major urban renewal scheme is proposed. A *ZAC* is usually initiated by the *'collectivité locale'* but its planning and implementation are often delegated to a *Société d'Economie Mixte (SEM)* (see below) or an *HLM* (social housing) organisation.

The objective of the *ZAC* is to create *'terrain a bâtir'* i.e. land ready for development. The *ZAC* provides a detailed plan for coordinated development; withdraws the area from the *POS* (the statutory local plan) in order that a separate detailed plan may be prepared; and withdraws the area of development from the obligations to pay *Taxe Locale d'Equipement (TLE)* (a tax that helps pay for local infrastructure and services) in order to substitute a higher level of developer participation in the financing of local infrastructure and communal facilities. The *ZAC* is frequently used in major redevelopment situations. For example the Ville de Paris has created some fifteen *ZAC* including the major derelict sites of the former Citroen factory in the fifteenth *arrondissement* and Les Halles in the heart of the city.

Compulsory purchase (expropriation) is not commonly used in French urban policy. Most purchases by local authorities are carried out by negotiation. This process is backed up by the *'droit de pré-emption'* which gives the local council the right of first refusal on any property that comes on to the market in certain specified areas.

There are two types of pre-emption zones: in the first case the local authority may designate part of its urban area as a *Zone d'Intervention Foncière (ZIF)* wherein pre-emption may take place at market prices. This right can only be exercised for the purposes of providing land for social housing or community purposes. Thus it is very appropriate for intervention in the re-use of abandoned premises and has often been used for that purpose. Although many urban areas are designated as *ZIF* the actual level of purchases is usually very low and confined to strategic interventions to facilitate regeneration. For example in one recent year the Ville de Paris made offers on about 1,300 eligible premises (about 14 per cent) but only bought about twenty properties. There are two reasons for this: firstly the right of pre-emption is as much a tool of coercion as it is for land purchase, with the local authority using the threat of acquisition (at a relatively low price) to persuade an owner to sell to another agency (e.g. an *HLM*) for a slightly better but still reasonable price. Secondly, the actual level of purchases is limited by the relationship between the prevailing price of land in the area and the resources of the local authority, so in Paris, where local authority resources are stretched and property values are high, particularly in the west of the city, purchases are few.

A related policy instrument is the *Zone d'Aménagement Différé (ZAD)*. This is an area where a local authority anticipates substantial development or redevelopment activity and thinks it desirable to control the land market and perhaps to acquire certain parcels of land. Unlike a *ZIF* which is permanent, a *ZAD* is designated for fourteen years and gives the local authority a right of pre-emption but at a price that is determined by the land *use* in existence (NB: not the land *value*) one year before designation. Also within a *ZAD* owners may require the local authority to purchase their property within one year of designation. Although *ZAD* are used less frequently today than formerly they have been found useful in large scale rehabilitation and redevelopment projects where the purchase of key sites and the control of land speculation may be critical.

Both *ZIF* and *ZAD* procedures rely upon the willingness of owners to sell their property. Thus a major feature of French redevelopment schemes is a dependence upon long negotiations and the need for flexibility in the planning of schemes to accommodate the desires of individual owners.

A further distinctive feature of French redevelopment/reclamation activity is the close relationship that exists between the public and private sectors. While it is unusual for local authorities to carry out much of the actual redevelopment work themselves, it is equally unusual for matters to be left in the hands of private developers or central government agencies. The favoured French solution makes use of a *Société d'Economie Mixte (SEM)*. These are joint-stock companies whose ownership is shared between public and private sector interests with the public sector, usually in the form of one or more local authorities, maintaining a controlling interest of between 51 per cent and 65 per cent of the shares. *SEM* are used by local authorities to carry out a wide variety of planning, development and management work on their behalf.

For many local authorities the advantage of using a *SEM* to carry out works are that: they bring in expertise that the authority may not possess; they are geared up to the quick and efficient decision taking required in development projects; they can attract private sector capital; and, theoretically, they permit the local authority to maintain democratic control over a project while keeping at arm's length from day to day involvement. The *SEM* may be given powers by the local authority to act as its agent, for example in a *ZAC*, from drawing up the *Plan d'Aménagement du Zone* to land assembly, development supervising, and disposal or management of completed buildings.

One of the most spectacular examples of French urban renewal can be found at Les Halles in the heart of Paris. When the former wholesale markets moved out to the suburbs the city found itself faced with a large vacant block of land right in the very heart of the city. The response was to designate the area a *ZAC* and to establish a *Société d'Economie Mixte SEMAH* created in 1969 with 51 per cent capital from the Ville de Paris, 25 per cent from the state and 24 per cent from other investors). *SEMAH* was given responsibility for the entire redevelopment process from strategy, through feasibility studies to the management and coordination of the redevelopment works.

The initial proposals for the area reflected the financial value of the site and the economic centrality of the area. During the seventies, in response to growing pressure for conservation and greater sensitivity in planning, the proposals changed and, in their final form emerged as a relatively moderate density development of low-rise building and a considerable amount of open space and community facilities carefully inserted into the context provided by the existing surroundings.

The final scheme included over 800 units of social housing, community and cultural facilities and an elaborately landscaped garden as well as hotels and substantial amounts of retailing and office floorspace.

The remarkable features of Les Halles are its scale (the *1978* estimate of total cost of redevelopment was FF1,612 million); its architectural style and ingenuity of design; and, the way *SEMAH* and the Ville de Paris eventually resisted the strong self-interested pressures coming from the development industry, in favour of a wider view of the Paris economy which, on the one hand recognised the area's potential contribution to the tourism economy and yet on the other hand acknowledged the importance of maintaining subsidised social housing and community facilities in the heart of the conurbation.

Housing renewal

Since the late seventies Government policy towards housing and housing renovation has been characterised by a series of features: (i) a decline in Central Government current expenditure on housing as subsidies are gradually withdrawn from local authorities and council housing rents pushed towards market levels; (ii) a decline in Central Government capital expenditure as the State withdraws from mainstream housing provision, contenting itself with a residual role and leaving most aspects of housing investment to the vagaries of the free market; (iii) within this decline in capital expenditure an increase in the proportions devoted to housing renovation and improvement. These changes are illustrated in Table 7.3.

Table 7.3 The State and Housing Performance in Britain
Central Government Expenditure on Housing: (£million at current prices)

	78/79	79/80	80/81	81/82	82/83	83/84	84/85	85/86	86/87	87/88
Total Current Expenditure	1922	2499	2918	2426	2016	1731	1692	1710	1701	1717
Total Capital Expenditure	3024	3602	3386	2328	2337	3013	2951	2538	2167	1826
of which										
Improvement £m	109	156	178	269	561	1214	982	641	584	542
Grants (as a % of capital expenditure)	3.6	4.3	5.6	11.5	24.0	40.2	33.3	25.3	26.9	29.7

Source: Housing and Construction Statistics

Looking back over housing renewal policy in recent years certain general comments can be made.

(i) Housing improvements policy has suffered erratic and generally insufficient resourcing since 1974.
(ii) Recently questions have been raised about the degree of spatial concentration remaining in housing obsolescence today and the consequent continuing validity of the area based approach to policy.
(iii) There have been questions raised about the dominance of improvement over clearance policy and calls for a return to a higher level of clearance, albeit gradual renewal with community involvement.
(iv) There are questions about the quality of improvement and repair work, especially in the early years of area improvement.
(v) There are questions about the long run implications of improvement policy: will these dwellings continue to meet housing aspirations and needs into the 1990s and beyond the turn of the century? Can these dwellings be re-improved or is there an impending clearance crisis about to emerge?

Recently the government has responded to some of these points by announcing a new approach to area improvement from 1990. Under the proposals local authorities will be able to designate a 'Renewal Area' which will be larger in area than the present GIAs and HAAs. Within these areas there will be scope for area improvement schemes, housing improvement and the refurbishment or redevelopment of commercial premises. It is the government's policy that home improvement grants should be means tested and that there should be opportunities for private sector investment in the improvement process. The implementation of area renewal under these new proposals is therefore likely to require substantial cooperation between local authorities, householders, local businesses and the private sector funding agencies, be they building societies, developers or whoever. One further consequence of broadening the areas is that they are now likely to include the selective demolition of property, either for unfitness reasons or to assist with area improvement proposals. Unless such demolitions are tightly planned there could be a significant blighting effect on private investment which might undo much of the benefit of earlier improvement programmes.

7.3 Intervention Through New Agencies and Organisational Changes

Urban Development Corporations

The legal basis for Urban Development Corporations (UDCs) was established by the 1980 Local Government, Planning and Land Act. Under this legislation the Secretary of State for the Environment may designate areas and set up UDCs to promote regeneration through the reclamation of derelict land and buildings, encouraging industrial and commercial development and ensuring the provision of social facilities and housing. Essentially the UDCs are an urban manifestation of the New Town Development Corporations pioneered by the Labour government nearly forty years earlier. One of the major criticisms of each type of agency has been their lack of local accountability. In the case of new towns this was justified on the basis that these developments were to help re-house urban slum dwellers and that such action was not to be frustrated by the rural middle classes. In the case of UDCs the government's argument was based upon the desirability of achieving speed and efficiency in decision making and an unspoken distrust of the local (usually Labour) councils.

UDCs have power to acquire, manage and dispose of land, to carry out reclamation works and to provide roads and other infrastructure. They can also, most controversially, be given substantial powers of development control which effectively makes them the planning authority for their area. UDCs are run by a board appointed by the Secretary of State. Although such boards usually include some local representatives this is seen by many as a poor substitute for local democracy.

The first UDCs were established in 1981 in the London and Liverpool Docklands: the London Docklands Development Corporation (LDDC) (2,000 hectares) and the Merseyside Development Corporation (350 hectares), in very different economic circumstances. Both UDCs were imposed by the government upon areas that already had their own plans for regeneration but not the vast central government funds proposed and were unanimously opposed to the loss of local control over redevelopment. The LDDC area comprised derelict docklands and large run down areas of the inner city (especially on the Isle of Dogs) stretching from the very edge of the buoyant City of London towards more depressed North Woolwich and Beckton some six miles to the east. The Merseyside area by contrast consisted almost entirely of derelict docklands with virtually no indigenous population and a very weak local economy.

Subsequently, in 1987 and 1988 further UDCs have been established in the West Midlands, Trafford Park, Central Manchester, Tyne and Wear, Teeside, Cardiff Bay, Leeds, Bristol and Sheffield.

Urban Development Corporations are financed direct from central government, as were the former new towns. Total resources going to UDCs have risen from £62m in 1982/83 (22 per cent of DOE 'urban' expenditure) to £210m in 1989/90 (42 per cent of DOE 'urban' expenditure). Indeed it is this bending of the DOE urban renewal programme and this concentration on a few 'privileged' areas that is a further source of substantial criticism. By contrast the Urban Programme has fallen from 56 per cent of expenditure to only 42 per cent and derelict land reclamation from 22 per cent to 15 per cent over the same period. (Government Expenditure Plans 1988/89–1990/91.)

Over the years since their inception both the LDDC and the MDC have seen impressive developments take place within their designated areas.

> Since 1981, over 3500ha of derelict land has been reclaimed in the LDDC area and £360m of public money has generated private sector commitment of over £2,200m, of which about 40 per cent has gone on private housing development. In the MDC area £140m of public investment has helped to reclaim 97 hectares for residential and commercial development and 48 hectares for recreation and public open space; and to refurbish 135,000 square metres for housing and commercial uses, including the historic Albert Dock restoration. The MDC has also created 1,160 jobs since 1981 and 94 per cent of its contracts have been let to firms in the Merseyside area.
>
> (DOE, 1988, pp. 52–53)

However this progress has been achieved against continuing criticism of UDCs and the way they operate. There have been substantial specific errors or problems: in Liverpool the 1984 International Garden Festival (a tremendous success in itself and a credit to its organisers) went ahead in the knowledge that no proper arrangements had been made for the continued funding and operation of the gardens after the end of the festival that year. Today the site lies closed and mothballed: an asset lost to local people and a monument to the consequences of lack of forward planning. In London there has been an almost complete failure to relate land use developments to their transportation consequences and needs resulting in a situation in which the new Docklands Light Railway is, at times, operating above capacity only a couple of years after its opening, and in which the local road network is solidly congested for many hours each weekday.

Very little official analysis of the performance of UDCs had been carried out until the National Audit Office (NAO) published a study in 1988. The NAO criticised the DOE for not clearly defining the meaning of 'regeneration' and therefore inadequately stating the objectives to which the UDCs should be working. They also made further detailed criticisms of DOE monitoring and control arrangements. With regard to the LDDC they made the important observations that:

> The LDDC's success partly reflects its proximity to central London and deregulation of the City. These special advantages do not detract from the LDDC's performance but they mean that its achievements will not readily be repeated elsewhere...The LDDC has met opposition and lack of cooperation from local authorities....There have also been difficulties in coordination with education and health authorities.
>
> (National Audit Office, 1988b, p.3)

With regard to the Merseyside Development Corporation, the National Audit Office took the view that while it had achieved a few notable successes within a difficult economic climate, there had been little progress in job creation and much of its marketing had lacked focus or drive.

Taking a wider view the NAO felt that the DOE should have more regard to the economic potential of sites in the selection of future areas for UDC treatment. They also emphasised the need for better corporate planning and management and improvements in the coordination of the work of the UDCs with the investment programme of other agencies and government departments (i.e. more planning is needed!).

More strident criticism of the organisation and performance of UDCs comes from a report published by the London Docklands Consultative Committee. In their view:

> Market, or property led, solutions cannot provide the answer to inner city problems: UDCs lead to islands of prosperity being developed in a sea of disadvantage and deprivation.
>
> (Docklands Consultative Committee, 1988, p.37)

They also consider that local authority planning policies have been overridden; that in spite of local representation on the Board the LDDC is neither accountable nor responsive to local opinion; that in compulsorily acquiring premises for property development the LDDC has actually destroyed jobs in the area; that had the local authorities had the equivalent level of resources they could have invested more wisely and to better local effect; and that the LDDC has ignored many social objectives, such as equal opportunities, in the rush for 'development at any cost'.

So far the MDC has escaped much of this kind of criticism but this has more to do with the nature and scale of its operations hitherto (behind the Dock wall) than its mode of operating. Recently the MDC boundary has been extended and now includes large areas of existing population and employment. It is to be hoped that the MDC will learn the lessons of the importance of local community participation and more humble cooperation with local agencies which has so far been so evidently lacking in the London case. Figure 7.2 illustrates developments in the Liverpool Docklands.

Figure 7.2 Albert Dock: the jewel in the crown of the Merseyside Development Corporation

A good example of one of the newer generation of urban development corporations can be found in Manchester where the Central Manchester Development Corporation (CMDC) was established in 1988 to lead the redevelopment and regeneration of a 187 hectare strip of land on the southern periphery of Manchester city centre. The area bounds the Bridgewater and Ashton canals and the Deansgate–Piccadilly railway line, and contains a mixture of older industrial uses, workshops, warehousing, secondary offices and shopping together with a large amount of vacant and derelict land and buildings.

In recent years central Manchester has experienced an upswing in demand for commercial floorspace which has been matched by institutional investment in supply. There has also been a remarkable increase in tourism and leisure related developments.

The CMDC has established a development strategy aimed at:

(i) stimulating private investment, especially through the use of City Grant to promote developments that will extend developers confidence from the city centre into this more marginal area;
(ii) comprehensive improvements to the local physical environment including the canalsides, the encouragement of high quality design and, interestingly, enhancement of the five points (or ' gateways') where major radial roads into the city centre pass through the area;
(iii) a programme of land assembly so as to assist in the provision of key development sites and social housing.

In all, some £760m private sector money is currently proposed for investment in the area over the projected five year lifespan of the CMDC. In addition to development action within its boundaries, CMDC sees the building of the city's light rapid transit system, the rail link to Manchester Airport and completion of the city's inner relief road as making major contributions to the success of their redevelopment strategy.

Other public agencies and organisational changes

Merseyside Task Force

In the wake of the 1981 Toxteth riots the then Secretary of State for the Environment, Michael Hesletine, acquired the temporary title of 'Minister for Merseyside'. In order to provide better local back-up for his initiatives he established a branch office of the Manchester based DOE North-West Region in Merseyside. The function of this office, which became known as the Merseyside Task Force, was to establish priorities for government assistance, to coordinate the government response and encourage cooperation and joint working. The office continues to be responsible for the management of all DOE financial assistance to inner city regeneration in Merseyside. Recent examples of Task Force activities have included:

- providing £6.0m in grant aid and organisational support for the setting up of the 25-hectare Wavertree Technology Park, which now employs around 1,600 people in 40 enterprises;

- being instrumental in creating a consortium of major landowners to undertake a £40m transformation of 40 hectares of derelict land close to St Helens town centre;
- grant aiding the construction of the 'Anglican Cathedral College', student accommodation and conference facilities on a site to the west of the cathedral;
- a £0.84m grant to establish the Langrove Street Housing Cooperative and refurbish 30 homes.

Other Task Forces

A smaller and more streamlined version of the Task Force concept has more recently been applied in other areas. These aim to create jobs and training initiatives, and support local enterprise and environmental improvements through short-term tightly focussed crash programmes of government assistance. By 1987 Task Forces had been set up in sixteen inner city areas.

> The Task Force in Moss Side, Manchester, is typical. Like the other Task Forces, it has very limited funding – some £1m over a two year period – but in practice it can be argued that it is not primarily concerned with spending money. Its greatest potential innovation is as a local facilitating body through which to create networks between central government, the local council, the private sector, and the local community and thereby encourage the development of training and generation of local enterprise.
> (Robson, 1988, p.135)

Estate Action and the Priority Estates Project

In 1979 government concern about the very high levels of vacancy and local environment degradation on some council housing estates had risen to such a level that the DOE initiated the 'Priority Estates Project' to investigate and test ways of improving the management and maintenance of run-down council estates. After the initial experimental phase the Urban Housing Renewal Unit was established in order to provide a mainstream route for the channelling of government funds and 'good practice'. In 1987 the unit was renamed 'Estate Action' and is currently responsible for an annual budget of £140m (1988/89) encouraging tenant self-management schemes, cooperatives and other locally based management approaches. Finance is provided for the refurbishment of dwellings, enhancement of security, 'clean-up' campaigns, improved levels of maintenance and repair, and supporting local business initiatives.

Housing Action Trusts
Under the 1988 Housing Act the government has sought to tackle the very worst problem estates through the establishment of Housing Action Trusts. These would be single-minded organisations rather like small Urban Development Corporations, which would take over the ownership of run-down council estates, use central government funds for their improvement and then re-sell the property to new owners: probably housing associations or private landlords. In spite of the initial publicity and high profile given to this controversial proposal, opposition from local authorities, tenants and housing associations has been such that the government appears to be backing away from the concept. Very little money has been allocated for the scheme and it seems possible that only a very small number of 'experimental' schemes will now go ahead.

Private and charitable agencies

There are a number of private and charitable agencies now involved in various aspects of urban renewal. Here, three of the more active agencies are briefly described.

The Civic Trust Regeneration Unit
The Regeneration Unit was established in 1987 as an advisory unit offering architectural and planning expertise through: a rolling programme of regeneration studies and projects; an information point and helpline for those involved in regeneration; and advice on the setting up of development trusts and similar organisations. The work is financed from three main sources with approximately equal amounts coming from the Department of the Environment, fee income and sponsorship. Examples of its work include: working with other agencies on the preparation of an urban landscape and development strategy for Ilfracombe; the regeneration of Deptford High Street; working with Calderdale District Council on a rolling programme of regeneration projects; other work includes feasibility studies and campaigning work on regeneration issues.

The Phoenix Initiative
The Phoenix Initiative was established in 1986 to promote and facilitate investment and good practice in urban regeneration. Support was derived from a number of organisations with interests in urban renewal including the Building Employers' Confederation and the Building Societies Association. Its main activity is in encouraging the improvement of specific urban areas through the setting up of local organisations, including trusts or consortia of interested local agencies and firms. Thus the Phoenix Initiative

acts as a motivating, facilitating and advisory agency but would not itself participate in the renewal and development task.

Probe

Partnership Renewal of the Build Environment (Probe) is typical of a number of private sector consortia that have been established to tackle urban renewal projects. Probe comprises the Nationwide-Anglia and Halifax building societies and the construction company Lovell. Probe is, according to its own publicity:

> an action agency which takes a comprehensive approach to development initiatives in order to meet the policy objectives of public authorities, with private sector and building society funds.
>
> (Probe publicity handout, 1987)

So far Probe has been involved in a number of housing and industrial redevelopment and refurbishment schemes in the Black Country, Salford, Hertfordshire and elsewhere.

The Cooperative movement

One of the most interesting initiatives in recent years has been the growth of the housing cooperative movement. This has been particularly strong in Liverpool. One of the earliest developments was the Weller Street Cooperative. A small group of streets in the Toxteth area of the city were blighted by clearance proposals but no action. Maintenance of these private rented properties was poor and the environment deteriorating. In 1977 local residents formed a cooperative. With assistance from the Housing Corporation and the Liverpool-based Cooperative Development Services they purchased a plot of land near their former homes. A total of 61 dwellings were built on the site. Members of the cooperative participated fully in the design and management of the estate. By the mid-eighties there were over 30 cooperatives in Liverpool. About half were involved in new building on redevelopment sites and the remainder with housing rehabilitation projects. Between 1978 and 1985 over 2,000 homes were provided in the city through these cooperative renewal projects (CDS) (1987).

While the cooperative movement was backed by the then Liberal city administration and received financial and organisational support, the local elections of 1983 brought in a radical Labour administration which, against national party policy, was strongly opposed to the use of public funds to support cooperatives. Their argument was that they saw cooperatives as elitist, exclusive and discriminatory and that public funds should be channelled towards the provision of council owned housing to be allocated solely on the basis of proven housing need. During the mid-eighties the

Labour council froze further land sales to cooperatives and municipalised several schemes that were still in the early stages of development.

One of the most well known of the Liverpool cooperatives is the Eldonian Housing Coop (Figure 7.3). Created in October 1984 the primary objective of the group was to rehouse its members (145 families) from an area of slum housing and poor environment in a way that would keep the community together. Following the closure of the Tate & Lyle Sugar Refinery a competition was held for the re-use of the site. While not

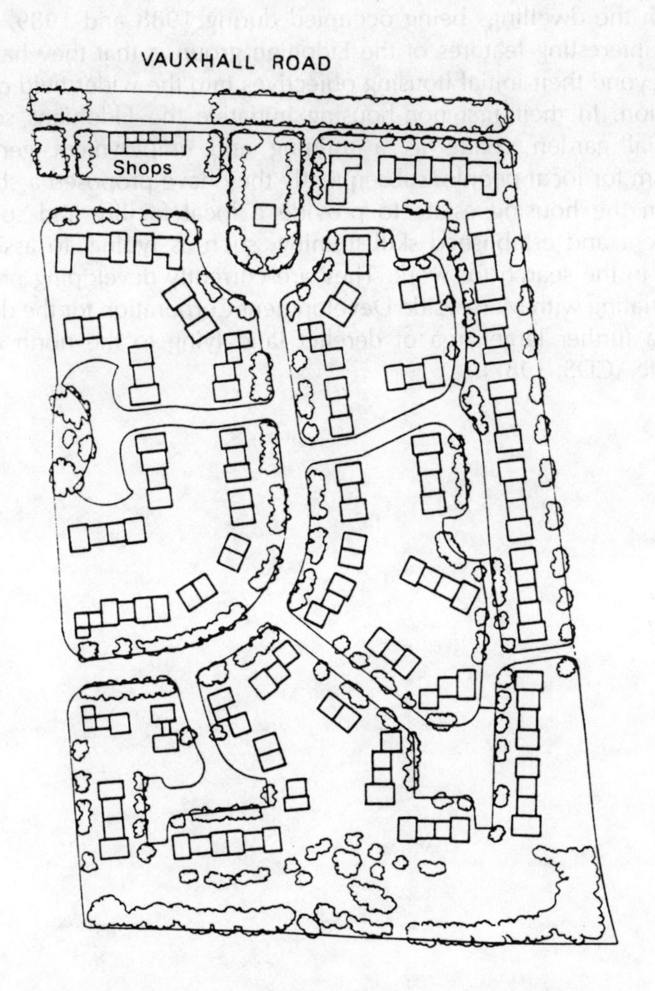

Figure 7.3 The Eldonian Housing Cooperative

winning the competition, the Eldonians (as they are known) sufficiently impressed English Estates (the landowners) with their proposals for community use that they agreed to offer them the land.

In May 1985 the Labour council refused planning permission for the Eldonians proposals but was over-ruled by the Secretary of State on appeal later in the year. In the autumn of 1986 the Department of the Environment announced funding of £6.6m for the reclamation of the derelict Tate & Lyle site and the building of 145 homes. Building work commenced in 1987 with the dwellings being occupied during 1988 and 1989. One of the most interesting features of the Eldonian group is that they have now moved beyond their initial housing objectives into the wider field of urban regeneration. In their first non-housing initiative the Eldonians set up a commercial garden centre as a training and employment generation mechanism for local people, subsequently they have proposed a shopping parade on the housing estate to provide a local facility and source of employment and established skill training, such as typing, to assist their members in the search for work. They are currently developing proposals and negotiating with Merseyside Development Corporation for the development of a further large area of derelict land lying to the north of their present site. (CDS, 1987).

8 Conclusions

This book has not been written around tightly defined hypotheses but has offered a critical discussion of a number of issues. It would therefore be inappropriate to come to any firm conclusions about the theory or practice of urban renewal. Nevertheless the discussion has raised a number of issues and areas of debate that are worth summarising here.

The first point is that urban renewal is clearly a multi-faceted and complex process and as such requires variety and subtlety in policy responses. Such responses need to be based upon a multi-disciplinary understanding of the social and economic forces affecting urban areas; the nature of government and organisations; and the physical nature of towns and cities. Both the study of urban renewal and the practice of intervention is a team effort based upon many skills.

It is a commonplace to note the tension and conflict that exists between labour and capital over the distribution of economic wealth. But changes in this relationship, with capital gaining strength in the 1930s and 1980s and labour making gains in the 1950s and 1960s has had significant impacts upon urban renewal: upon housing quality and the provision of community facilities and amenities. Another source of tension and conflict is that between the short run market economic pressures upon individual firms and consumers and the long run economic needs of capitalism and society in relation the urban system. There are long run economic requirements for physical urban infrastructure (buildings, transport systems and utilities) and public services (education and training, social and environmental control systems and so forth) that tend to be subject to cutbacks and neglect at times of crisis and yet are fundamental to the long run survival of urban areas. Britain could be said to be going through such a period now but there are signs of a gradual realisation of the consequences of this neglect of our urban fabric. In a wider sense also society is beginning to debate longer term objectives for urban life: personal health and safety in 'the healthy city', the quality of the environment, the use of energy, and the major issue of how to better use, accommodate and control the motor vehicle within urban areas.

While all urban renewal is subject to the same basic economic laws there is considerable variety in the social and economic conditions under which it takes place. These variations occur between and within urban areas. The renewal of parts of London and other towns in the South East takes place in a climate of sustained strong economic demand for urban space while the renewal of cities in more peripheral regions occurs in circumstances of relatively weak demand and frequently an oversupply of urban land and buildings. Within any one city demand and supply conditions will vary between different functional zones (city centre, inner city, suburban centres, etc.) and between different geographical districts. Government policy in the urban renewal field has not always reflected this variety: for example a recent evaluation of the Enterprise Zone policy considered that the level of subsidy given in EZs in the South East and East Midlands was above the level required to stimulate investment whereas, by implication, the level in other regions was too low.

Lessons from the historical evolution and current practice of urban renewal and from management theory stress the importance and advantages of local community participation in renewal decisions. Present government policy is confused on this matter, supporting participation in aspects of housing policy, tenant management schemes, housing cooperatives and local planning, and yet working against participation in urban development corporations, privatisation and in the allocation of subsidies such as City Grant.

There is a need to recognise the turbulent environment of urban renewal and to develop adaptive–organic organisations of intervention appropriate to this climate. To some extent the urban development corporations achieve this requirement but are flawed by their lack of community participation and local accountability. Some local authorities have set up enterprise boards, development agencies and area project teams that have some adaptive–organic characteristics. Housing cooperatives and other community based development organisations are other examples.

In recent years the main thrust of government policy has been to stimulate urban regeneration through various forms of subsidy and to maintain only tactical-level mechanisms of urban planning and control. The effect has tended to lead to piecemeal and uncoordinated urban renewal. The consequences of this approach are beginning to damage the efficiency with which resources are used and the effectiveness with which social objectives can be achieved. The results are most apparent in London where high levels of investment in urban renewal combined with a multiplicity of responsible agencies and no adequate overall control is leading to approval being given for huge investment proposals (such as Canary Wharf, Broadgate or the Kings Cross redevelopment) with little regard for their impact upon the local environment, employment and travel patterns, property investment and housing demand across the city and beyond. There is an urgent need for the development of a sound strategic level of urban renewal planning and control as well as the present range of tactical agencies working at the individual programme and project level. Such a policy instrument exists in the form of Structure Plans but they have been so discredited and undermined by other government policies in recent years that they command little respect today. In addition there needs to be an agency with appropriate powers and responsibilities to develop and enforce strategic policies. Both the policy instrument and the agency are missing from state intervention in urban renewal in Britain today.

It is increasingly difficult to achieve the kind of urban fabric that many people admire in certain historic cities: human scale and a sense of place, variety of building character and form within a basic unity of local style and materials. Economic demand is for increasingly large and standardised buildings which fit awkwardly into existing urban areas. While the planning system is quite effective in controlling land use it is relatively weak in the realms of design control and in the design and management of the public space between buildings. Many towns and cities choose to concentrate efforts upon the conservation of the best of the old, sometimes associated with tourism development policies. Few local planning authorities could claim to have a coherent urban design strategy for a complete town or city. If there is to be a significant improvement in the quality of urban design in urban renewal it is likely to require some fairly fundamental changes in attitudes and policy. For example, moves away from speculative developments towards more bespoke contracting might generate more concern for quality in building specification and design, including the immediate local environment, as well as energy efficiency and reductions in maintenance costs. Changes in the planning system to give planners more control over building form and the detailed design of areas (such as in the German *Bebauungsplan*) might generate debate about the nature and objectives of urban design as well as providing a mechanism for their achievement.

In summary there are a number of major questions and issues in urban renewal that deserve further consideration, investigation and debate. These include:

- the multi-disciplinary nature of the study of urban renewal, the need for variety in urban renewal policy and training for these activities;
- the long run goals of state and local community intervention in urban renewal, including questions of efficiency and equity, health and personal safety, the use and control of motor vehicles in urban areas, the effects of urban renewal upon energy use, pollution and environmental quality;
- the goals of urban design and ways in which they might be achieved;
- the need for a strategic level of urban renewal policy instruments and agencies;
- the importance and benefits of community participation in urban renewal;
- the need to learn from the experiences and examples of good practice found in other countries; especially in Western Europe.

References

Alexander C, Neis H, Anninou A and King I (1987) *A New Theory of Urban Design* Oxford University Press, Oxford
Ash J (1980) 'The Rise and Fall of High Rise Housing in England', in Ungerson C and Karn V (Eds) *The Consumer Experience of Housing* Gower, London
Ashton T S (1972) *An Economic History of England: the 18th Century* Methuen, London
Balchin P N and Kieve J L (1985) *Urban Land Economics* (3rd Ed.) Macmillan, London
Ball M (1988) *Rebuilding Construction: Economic Change and the British Construction Industry* Routledge, London
Bentley I, Alcock A, Murrain P, McGlynn S and Smith G (1985) *Responsive Environments* The Architectural Press, London
Burnett J (1978) *A Social History of Housing 1815–1970* Methuen, London
Burns T and Stalker G M (1966) *The Management of Innovation* Tavistock, London
Cameron G C, Monk S and Pearce B J (1988) *Vacant Urban Land: A Literature Review* DOE, London
CDS (1987) *Building Democracy: Housing Cooperatives on Merseyside* Cooperative Development Services, Liverpool
Cherry G E (1972) *Urban Change and Planning* G T Foulis, Henley on Thames
Community Forum (1987) *A Tale of Three Cities* Community Forum, Birmingham
Cooney (1974) 'High Flats in Local Authority Housing in England and Wales since 1945', in Sutcliffe A (Ed.) *Multi-Storey Living: The British Working Class Experience* Croom Helm, London
Couch C (1985) *Housing Conditions in Britain and Germany* Anglo-German Foundation, London
Couch C, Basnett M, Holmes T and Doward L (1989) Urban Vacancy in Liverpool in Grover R (Ed.) *Land and Property Development: New Directions* E&F N Spon, London
Couch C and Herson J (1986) The *Gründstückfond* Ruhr: A System for Managing Derelict Land *The Planner* Vol. 72 No. 9

Couch C and Morton R (1988) Regional Recession and the Structure of the UK Construction Industry *Production of the Built Environment* (BISS) Vol. 9, University College, London

Couch C and Wynne S (1986) *Housing Trends in Liverpool* Liverpool Council for Voluntary Service, Liverpool

Cullen G (1961) *Townscape* Architectural Press, London

Cyert R M and March J G (1963) *A Behavioural Theory of the Firm* Prentice Hall, New York

Deane P and Cole W A (1962) *British Economic Growth 1688–1959: Trends and Structures* Cambridge University Press, Cambridge

Docklands Consultative Committee (1988) *Urban Development Corporations: Six Years in London's Docklands* DCC, London

DOE (1972) *Environmental Design* Area Improvement Note No. 5 HMSO, London

DOE (1975) *Vacant Land* Liverpool Inner Area Study Report, DOE, London

DOE (1977a) *Unequal City* Final Report of the Birmingham Inner Area Study HMSO, London

DOE (1977b) *Inner London: Policies for Dispersal and Balance* Final Report of the Lambeth Inner Area Study HMSO, London

DOE (1977c) *Change or Decay* Final Report of the Liverpool Inner Area Study HMSO, London

DOE (1980) *Housing Requirements: A Guide to Information and Techniques* HMSO, London

DOE (1984) *Survey of Derelict Land in England 1982* HMSO, London

DOE (1987) *An Evaluation of the Enterprise Zone Experiment* Inner Cities Research Programme, HMSO, London

DOE (1988) *Urban Land Markets in the United Kingdom* HMSO, London

Donnelly D (1987) *Decentralisation: A Special Case of Change in British Local Government* School of Social Science, Liverpool Polytechnic

Dunleavy (1981) *Politics of Mass Housing in Britain 1945–1975* Clarendon, Oxford

Edwards M (1985) Planning and the Land Market: Problems, Prospects and Strategy in Ball M (Ed.) *Land Rent, Housing and Urban Planning: A European Perspective* Croom Helm: London

Erickson R A and Syms P M (1986) The Effects of Enterprise Zones on Local Property Markets *Regional Studies* Vol. 20 pp.1–14

Essex County Council (1973) *A Design Guide for Residential Areas* Essex County Council, Chelmsford

Fayol H (1949) *General and Industrial Management* Pitman, London

Feinstein C (1981) Capital Accumulation and the Industrial Revolution, in Floud R and McCloskey (Eds) *The Economic History of Britain since 1700 Vol.1. 1700–1860* Cambridge University Press, Cambridge

Fothergill S, Kitson M and Monk S (1985) *Urban Industrial Change* DOE/DTI Inner Cities Research Programme, HMSO, London

Gans H J (1962) *The Urban Villagers* Free Press, New York
Gans H J (1967) *The Levittowners* Allan Lane, London
Gaudie E (1974) *Cruel Habitations: A History of Working Class Housing 1780–1918* Allen & Unwin, London
Goddard J B & Champion A G (Eds) (1983) *The Urban and Regional Transformation of Britain* Methuen, London
Goodall B (1972) *The Economics of Urban Areas* Pergammon, Oxford
Hall P (1975) *Urban and Regional Planning* Penguin, Harmondsworth
Hall P and Hay D (1980) *Growth Centres in the European Urban System* Heinemann, London
Harvey J (1981) *The Economics of Real Property* Macmillan, London
Hillier B (1988) Against Enclosure, in Teymur N, Markus T and Wolley T (Eds) *Rehumanizing Housing* Butterworth, London
Hobsbawn E J (1968) *Industry and Empire* Penguin, Harmondsworth
Holcomb B (1984) Women in the Rebuilt Urban Environment: The United States Experience *Built Environment* Vol. 10 No. 1
Howard E (1985) *Garden Cities of Tomorrow* Attic Books, Eastbourne.
Jacobs J (1961) *The Death and Life of Great American Cities* Penguin, Harmondsworth
Jacobs J (1984) *Cities and the Wealth of Nations* Viking Press, New York
JURUE (1986a) *An Evaluation of Industrial and Commercial Improvement Areas* Inner Cities Research Programme, DOE, HMSO, London
JURUE (1986b) *Evaluation of Environmental Projects Funded under the Urban Programme* DOE Inner Cities Research Programme, HMSO, London
JURUE (1988) *Improving Urban Areas: Case Studies of Good Practice in Urban Regeneration* DOE, HMSO, London
Kast F E and Rosenzweig J E (1981) *Organisation and Management: A Systems and Contingency Approach* (3rd Ed.) McGraw Hill, New York
Keeble L (1952) *Principles and Practice of Town and Country Planning* The Estates Gazette, London
Knox P (1982) *Urban Social Geography* Longman, London
Lal R (1983) *The Rise and Fall of High Rise Housing* Department of Town and Country Planning, Liverpool Polytechnic
Lansley P, Sadler P and Webb T (1977) Managing for Success in the Building Industry *Building Technology and Management* Vol. 13 No.7 pp. 21–23.
Lawless P (1979) *Urban Deprivation and Government Initiative* Faber & Faber, London
Lever W F and Moore C (Eds) (1986) *The City in Transition* Clarendon Press, Oxford
Lickert R (1961) *New Patterns of Management* McGraw-Hill, Maidenhead
Lindblom C E (1959) The Science of Muddling Through *Public Administration Review* Vol. 19 pp. 79–88

Loney M (1983) *Community Against Government: The British Community Development Programme 1968–1978: A Study of Government Incompetence* Heinemann, London
Lynch K (1959) *The Image of the City* MIT Press, Cambridge MASS
MacDonald R (1989a) Liverpool North Docklands: The Potential for Urban Design and Industrial Regeneration *Urban Design Quarterly* No. 29
MacDonald R (1989b) The European Healthy Cities Project *Urban Design Quarterly* No. 30
Merseyside County Council (1975) *Stage One Report* Merseyside County Council, Liverpool
Merseyside County Council (1979) *Merseyside Structure Plan: Written Statement* Merseyside County Council, Liverpool
MHLG (1952) *The Density of Residential Areas* HMSO, London
MHLG (1958) *Flats and Houses* HMSO, London
MHLG (1962) *Residential Areas: Higher Densities* HMSO, London
MHLG (1965) *The Future of Development Plans* Report of the Planning Advisory Group, HMSO, London
MHLG (1966) *The Deeplish Study – Improvement Possibilities in a District of Rochdale* HMSO, London
MHLG (1970) *Development Plans: A Manual on Form and Content* HMSO, London
Midwinter E (1972) *Priority Education: An Account of the Liverpool Project* Penguin, Harmondsworth
Milner-Holland (1965) *Report of the Committee on Housing in Greater London* HMSO, London
Mintzberg H (1973) *The Nature of Managerial Work* Harper and Row, London
Morgan P and Nixon P (1988) *Building on the Past: An Overview of Conservation Policy* Derek, Wade & Waters, Preston
Nabarro R and Richards D (1979) *Wasteland: a Thames Television Report* Thames, London
National Audit Office (1988a) *Department of the Environment: Derelict Land Grant* HMSO, London
National Audit Office (1988b) *Department of the Environment: Urban Development Corporations* HMSO, London
National CDP (1976) *Whatever Happened to Council Housing* National Community Development Project
National CDP (1977) *Costs of Industrial Change* National Community Development Project
National CDP (1977) *Gilding the Ghetto: The State and the Poverty Experiments* National Community Development Project
Needleman L (1965) *The Economics of Housing* Staples Press, London
O'Connor J (1973) *The Fiscal Crisis of the State* St James Press, New York

Peters T J and Watermann R M (1982) *In Search of Excellence: Lessons from America's Best Run Companies* Harper and Row, London
Powell C G (1980) *An Economic History of the British Building Industry 1815–1979* Architectural Press, London
Pugh D S, Hickson, and Hinnings C R (1983) *Writers on Organisations* (3rd Ed.) Penguin, Harmondsworth
Reilly C and Aslan N J (1947) *Outline Plan for the County Borough of Birkenhead* Birkenhead County Borough Council
RICS (1983) *Listed Buildings and Conservation Areas* Surveyors Publications, London
Robson B (1988) *Those Inner Cities* Clarendon Press, Oxford
RTPI/CRE (1983) *Planning for a Multi-Racial Britain* Commission for Racial Equality, London
Sharp T (1968) *Town and Townscape* John Murray, London
Shore P (1976) *Inner Urban Policy* A Speech, Central Office of Information Press Notice, London
Simon H (1960) *The New Science of Management Decision* Harper and Row, London
Smith D (1976) *The Facts about Racial Disadvantage* Policy Studies Institute, London
Sutcliffe A (Ed.) (1974) *Multi-Storey Housing: The British Working Class Experience* Croom Helm, London
Thomas A (1983) Planning in Residential Conservation Areas *Progress in Planning* Vol. 20, pp. 173–256
Tym R and Partners (1984) *Monitoring Enterprise Zones: Year Three Report* DOE, London
Tym R and Partners and Land Use Consultants (1987) *Evaluation of Derelict Land Grant Schemes* HMSO, London
Wekerle G R (1988) From Refuge to Service Center: Neighbourhoods that Support Women, in van Vliet W (Ed.) *Women, Housing and Community* Avebury, Aldershot
Worskett R (1969) *The Character of Towns* The Architectural Press, London
Young M and Wilmott P (1957) *Family and Kinship in East London* Routledge and Kegan Paul, London

Index